CALIFORNIA NATURAL HISTORY GUIDES

INTRODUCTION TO THE CALIFORNIA CONDOR

California Natural History Guides

Phyllis M. Faber and Bruce M. Pavlik, General Editors

Introduction to the

CALIFORNIA CONDOR

Noel F. R. Snyder
Helen A. Snyder

UNIVERSITY OF CALIFORNIA PRESS
Berkeley   Los Angeles   London

To all those who have striven
to conserve the California Condor

California Natural History Guides No. 81

University of California Press
Berkeley and Los Angeles, California

University of California Press, Ltd.
London, England

Library of Congress Cataloging-in-Publication Data

Snyder, Noel F. R.
    Introduction to the California condor / Noel F. R. Snyder, Helen A. Snyder.
        p.   cm. — (California natural history guides ; 81)
    Includes bibliographical references and index.
    ISBN 0–520-23924-5 (cloth : alk. paper) — ISBN 0–520-24256-4 (pbk. : alk.
paper)
    1. California condor.   2. California condor—Conservation.   I. Snyder, Helen,
1942–.   II. Title.   III. Series.

QL696.C53S59   2005
598.99'2—dc22   2004006945

Manufactured in China
10   09   08   07   06   05
10   9   8   7   6   5   4   3   2   1

The paper used in this publication meets the minimum requirements of
ANSI/NISO Z39.48–1992 (R 1997) (Permanence of Paper). ♾

Cover photograph: An adult California Condor on a roost tree atop Mount Pinos
in 1970. Photograph by Noel Snyder.

Chapter 1 opening photograph: An adult California Condor sails along a
chaparral-covered ridgeline near its nest in Santa Barbara County in 1982.

Chapter 10 opening photograph: Four California Condors leave their roost along
the Piru River heading toward the foraging grounds in the San Joaquin Valley in
early 1980.

The publisher gratefully acknowledges the generous
contributions to this book provided by

the Moore Family Foundation
Richard & Rhoda Goldman Fund
and
the General Endowment Fund of the
University of California Press Associates.

CONTENTS

Preface   ix

Acknowledgments   xi

1. GIANT SCAVENGERS   **1**

Important Characteristics of the California Condor   7

Scavenging as a Lifestyle   16

Relationship of Body Size and Bill Shape to Diet   20

Other Specific Adaptations for Scavenging   22

2. FOOD AND MOVEMENTS   **33**

Diet Diversity   34

Quantitative Food Needs   37

Finding and Competing for Food   38

Behavior While Ingesting Food   46

Foraging Regions   53

Daily Movements of Breeders and Nonbreeders   57

Movements Related to Seasonal Changes in
Food Availability   58

Changes in Food Availability in Historical Times   60

3. BREEDING BIOLOGY   **63**

Courtship and Pair Formation   66

Nest Sites   71

Nest Investigations   73

Egg Laying and Incubation   75

	Hatching and the Nestling Period	77
	Fledging	83
	Natural Enemies of Breeding Condors	86
4.	**THE HISTORIC DECLINE**	**93**
	Early Records	94
	First Population Estimates	97
	Counts of the 1980s and a General Assessment	105
	Condor Numbers around 1900	106
5.	**WHAT CAUSED THE HISTORICAL DECLINE? EARLY HYPOTHESES**	**111**
	Shooting	114
	Poisoning	116
	Food Scarcity	122
	Human Disturbance of Nesting Areas	125
	DDE Contamination	127
	Collisions	128
	Calcium Stress	130
	Habitat Loss	131
	Other Miscellaneous Stresses	133
6.	**STUDIES OF THE DECLINE IN THE 1980S**	**137**
	Censusing Efforts and Determination of Mortality Rates	139
	Reproductive Studies of the 1980s	144
	Mortality Studies of the 1980s and Later Years	156
7.	**HISTORICAL CONSERVATION EFFORTS**	**171**
	Habitat Protection	172
	Other Historic Conservation Measures	176

Miscellaneous Conservation Efforts of the 1970s
and Early 1980s   178
Summary of Early Conservation Efforts   179

8. CAPTIVE BREEDING   **181**

Early Opposition to Captive Breeding   182
Reversing the Opposition to Captive Breeding   187
Establishment of a Captive Flock   188
Multiple-Clutching   192
An Early-Release Proposal and the Crisis of 1985   196
Reproductive Performance of the Captive Flock   199
Captive Breeding and Rearing Procedures   200
Genetic Concerns   204

9. RELEASES INTO THE WILD   **209**

A Rationale for Releases   211
First Releases of Andean and California Condors   214
A Conference on Behavioral Problems   219
Additional Problems   222
Summary of Progress   224

**10. CONDOR CONSERVATION
       IN A CHANGING WORLD**   **227**

Solutions to the Lead Problem   228
Quality versus Quantity in Release Strategies   233
Unforeseen Problems   238

Timeline of Important Habitat Protection Actions   241
References and Further Reading   245
Art Credits   257
Index   259

PREFACE

The present book is our third general offering on the biology and conservation of the California Condor and complements a number of technical scientific papers we have published on various aspects of this species' biology, based primarily on our field studies of the historic wild Condor population between 1980 and 1986. The focus in this volume is on a basic understanding of the natural history of the species and how this relates to efforts to conserve the species. Our previous general accounts of the species in 1989 and 2000 gave considerably more detail on the political complexities involved in Condor conservation and on the history of Condor research efforts, aimed more toward specialists in ornithological studies and conservation strategies. Those seeking more detailed information on such matters may wish to consult these earlier publications.

Over the years, a tremendous number of published accounts of the California Condor have appeared, both for scientists and for the lay public, and the most important of these are referenced in the References and Further Reading. It is clear that humanity has had an enduring fascination for the Condor. Evidence from studies of Native American societies indicates that this fascination extends far back into prehistoric times. Of previous publications on the species, perhaps the most pivotal have been the scientific monograph entitled *The California Condor,* published by Carl Koford in 1953, and a delightful general account of the species by Ian McMillan in 1968 entitled *Man and the California Condor,* but these are only two of many dozens of books and articles. The Condor has always been a highly controversial species, and no two accounts cover the same ground or present the same orientation on conservation priorities. The present account summarizes earlier accounts and studies but also presents our personal view on the biology and conservation of the species.

ACKNOWLEDGMENTS

Few species have received the depth and breadth of conservation attention that has characterized the efforts to preserve the California Condor, and any book attempting to review the biology and conservation of this species stands in major debt to the contributions of a large number of dedicated individuals and organizations. It is a major credit to the efforts of all these people and organizations that the species still exists, despite a strong professional consensus for more than a century that its disappearance was inevitable.

Among the organizations that have significantly aided the welfare of the species, we make special mention of the contributions of the U.S. Fish and Wildlife Service, the National Audubon Society, the U.S. Forest Service, the California Department of Fish and Game, the Bureau of Land Management, the Zoological Society of San Diego, the Los Angeles Zoo, the Peregrine Fund, the Ventana Wilderness Society, the University of California at Berkeley, California Polytechnic University at San Luis Obispo, the Santa Barbara Natural History Museum, the Western Foundation of Vertebrate Zoology, Hawk Mountain Sanctuary Association, the Illinois Natural History Survey, the Santa Cruz Predatory Bird Research Group, various California chapters of the Audubon Society, the Condor Survival Fund, Hearst Foundation, and Van Nuys Charities. Without the efforts of these organizations, the species would indeed have disappeared by now.

Among historic individuals contributing to the welfare of the Condor, we wish especially to acknowledge the efforts and publications of Charles Bendire, John Borneman, Dean Carrier, James Cooper, William Leon Dawson, two Robert Eastons, Ray Erickson, William Finley, Joseph Grinnell, Harry Harris, Carl Koford, Robert Mallete, Ian and Eben McMillan, Alden Miller, Cyril Robinson, Fred Sibley, Dick Smith, Frank Todd, and Sanford Wilbur. These people were crucial in the first research efforts on the species through to the 1970s and in early conservation efforts, such as the establishment of the Sisquoc and Sespe Condor Sanctuaries, establishment of the October Survey, and establishment of the federal Endangered Species Acts.

Nevertheless, our most heartfelt acknowledgments go to the many individuals involved in the program of the 1980s, when research efforts were most intensive and desperate last-minute efforts to establish a captive population took place. Without detailing their individual contributions, we wish to at least mention the following personnel who participated in the diverse operations of that period: Jack Allen, Louis Andaloro, Marilyn Anderson, Victor Apanius, Lee Aulman, Bruce Barbour, Pete Bloom, Gene Bourassa, Bill Burke, Brad Bush, Jim Carpenter, Mike Clark, Dave Clendenen, Bill Cochran, Cathleen Cox, Mike Cunningham, Jim Dalton, Scott Derrickson, Phil Ensley, Gary Falxa, Abel Galletti, Ben Gonzalez, Jesse Grantham, Jan and Hank Hamber, Dave Harlow, Leon Hecht, Marcia Hobbs, Jack Ingram, Don Janssen, Eric Johnson, Ron Jurek, Brian Kahn, Lloyd Kiff, Steve Kimple, Cyndi Kuehler, Gary Kuehn, Arlene Kumamoto, Tom Leckey, Dave Ledig, Bill Lehman, Don Lindburg, Mike Loomis, Cindy McConathy, Charlie McGlaughlin, Vicky Meretsky, Barbara Nichols, John Ogden, Butch Olendorff, Jim Oosterhuis, Randy Perry, Sandy Pletschet, Rob Ramey, Larry Riopelle, Bob Risebrough, Art Risser, John Roach, Bay Roberts, John Rogers, John Roser, Joe Russin, Greg Sanders, John Schmitt, Amy Shima, Dick Smith, Sandy Sprunt, Meg Stein, Don Sterner, Cynthia Stringfield,

Cindy Studer, Warren Thomas, Russell Thorstrom, Rick Throckmorton, Bill Toone, Rebecca Usnik, Jared Verner, Mike Wallace, Stan Wiemeyer, Jim Wiley, Pat Witman, Buck Woods, and Teresa Woods. These people worked exceptionally hard and were responsible for a major part of the achievements of that period.

For the late 1980s through the early 2000s we wish to mention the contributions of Gregg Austin, Mike Barth, Dave Clendenen, Shawn Farry, Dave Ledig, Allan Mee, Robert Mesta, Bruce Palmer, Hank Pattee, Kelly Sorenson, Mark Vekasy, Mike Wallace, Marc Weitzel, and Jim Wiley in leading various aspects of release efforts with the species.

For use of various historical photographs in the book, we are grateful to the Museum of Vertebrate Zoology at Berkeley, the U.S. Fish and Wildlife Service, the U.S. Forest Service, the Zoological Society of San Diego, the Western Foundation of Vertebrate Zoology, Phil Ensley, and Eric Johnson. Many of the photos presented were taken by Dave Clendenen of the USFWS, who also aided in reviewing parts of the manuscript, as did Vicky Meretsky and Walter Koenig. Dave Utterback provided crucial assistance in assembly of photographic materials. John Schmitt kindly provided two plates illustrating aspects of Condor biology.

WE LIVE IN turbulent times. As the twenty-first century begins to unfold, many natural environments of the earth are under major assault by forces ranging from climate change and chemical contamination to outright habitat obliteration. Numerous wildlife species will likely be lost in the years just ahead, and no one knows for sure whether a sustainable global biosphere can persist in the face of continuing human population growth and accelerating resource exploitation. What the world may look like in another century is something that few would dare to predict.

Yet despite the increasing pace of change and the many uncertainties that lie ahead, we still share the planet with a number of ancient living giants. Some of the largest life-forms that ever evolved, including the largest animal and plant species of all times, are still in existence today, dwarfing their surroundings and inspiring our respect by their very immensity. To be sure, certain of these huge creatures now face grave threats to their continued existence, and gigantism in itself can lead to a heightened vulnerability to extinction. Nevertheless, some of today's giants seem reasonably secure in their immediate prospects and may well prove to be among the survivors of the present ecological crisis. We need to recognize that extinctions can result from a great variety of negative forces and that the species vanishing in recent times have differed greatly in size and other characteristics. It is unclear to what extent gigantism per se should be considered a major cause of the disappearance of species.

Curiously, and perhaps by chance alone, a surprising number of the world's largest creatures occur in California. The Blue Whale *(Balaenoptera musculus)*, the largest animal to have ever existed, is regularly seen in California's coastal waters. The largest known plant species of all time, the giant sequoia *(Sequoiadendron giganteum),* thrives in the western Sierra Nevadas of the central part of the state. Appropriately enough, a third California giant that is still extant (albeit barely) once occasionally fed upon beached Blue Whales and occasionally nested in cavelike cavities in the trunks of giant sequoias (pl. 1). This species, the California Condor *(Gymnogyps californianus)*, was very nearly lost two decades ago and now survives primarily in captivity. Its numbers have been increasing, however, and initial attempts at reestablishing wild populations are underway.

Plate 1. One active California Condor nest of 1984 was 100 feet up in a cavity of a giant sequoia, but less than halfway to the top of the tree.

As a widely revered and truly awe-inspiring symbol of wilderness, the California Condor is the focal species of this book and has been the subject of intensive research and conservation efforts for many decades. We were personally involved in these efforts during the 1980s, especially the research to determine why the species was declining and the struggle to establish a viable captive population. In an earlier book published by Academic Press in 2000 — *The California Condor: A Saga of Natural History and Conservation* — we described many of the detailed results of those efforts. Here we present a more general introduction to the natural history and conservation of the California Condor, both to promote basic understanding of the nature of the species and to promote an appreciation of the difficulties ahead in ensuring its survival in an increasingly unstable world. As a species that has somehow persisted to the present, despite predictions of imminent extinction going back for more than a century, the Condor just may have enough resilience to continue to survive, although like ourselves it now faces the most ominous era of ecological change our planet has seen in many millions of years.

In early historical times the California Condor was widely distributed along the west coast of North America, from present-day British Columbia in Canada to the mountains of northern Baja California in Mexico (map 1). It was most abundant in regions near the Pacific Ocean in California and Oregon but also occurred irregularly as far eastward as what is now Alberta, Idaho, Utah, Arizona, and perhaps even Colorado. Even earlier, during the Pleistocene, fossils indicate that the Condor ranged as far eastward as present-day Texas, Florida, and New York (map 2), and although paleontological proof is not yet in hand, it is likely that the species may also have once occupied many other eastern, midwestern, and western localities. Evidently the Condor was once found throughout much of North America, even if the exact limits of its former range may never be fully established.

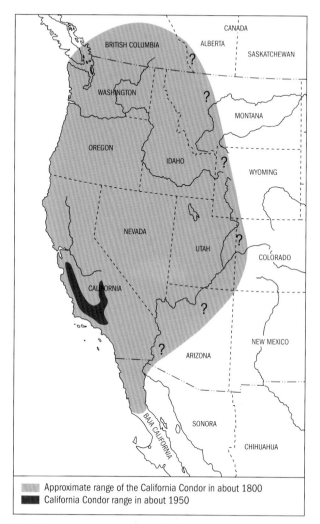

Approximate range of the California Condor in about 1800
California Condor range in about 1950

Map 1. The approximate range of the California Condor in early historic times (tan area; eastern boundary known only imperfectly) had shrunk enormously by the mid-twentieth century (red area). Solid lines represent coasts and rivers.

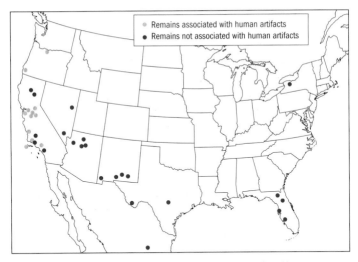

Map 2. Prehistoric California Condor remains have been found in many regions of the United States as well as in northern Mexico.

The Condor began to retreat from its full former range at a very early date, and by the end of the Pleistocene, roughly 10,000 years ago, the species had apparently already disappeared from all regions east of the Rocky Mountains. Very likely this early range contraction occurred because many of the species available as food for the Condor became extinct by the end of the Pleistocene. Some of the better-known species that disappeared were mammoths, sabre-toothed cats, and giant ground sloths (*Mammuthus*, *Smilodon*, and *Megatherium* spp.), but there were many others. Nevertheless, the Condor was among the survivors of the Pleistocene in the most westerly portions of North America and continued to occupy a wide range along the Pacific Ocean up until early historical times. It is unclear whether the population declined in any major way during the millennia immediately preceding the discovery of California by Europeans.

What is clear, however, is that the range of the species again began to contract steadily with the arrival of Europeans in the late eighteenth century. And in more recent times, the Condor became renowned as a symbol of disappearing wilderness and wildlife. By the start of the twentieth century, the species was limited almost exclusively to the southern half of California and northern Baja California; by the mid-twentieth century it was entirely confined to a horseshoe-shaped range in southern California surrounding the southern Central Valley. Here in its last bastion it continued to dwindle until 1987, when the very last wild bird was trapped into captivity.

The last few decades of existence of a wild California Condor population in the mid-twentieth century were a period of intense and bitter controversy regarding the fate of the species. Debate was continuous over the causes of the species' decline and the best methods to prevent the species' extinction, and despite all efforts to resolve these issues, the decline continued. Unfortunately, the efforts to understand and conserve the species were severely handicapped by the wide-ranging habits of individual Condors, difficulties in identifying individuals, and difficulties for researchers in traversing the rugged mountainous terrain occupied by the species. How knowledge of the causes of decline evolved through time, how the species was rescued by captive breeding in the late 1980s and 1990s, and how efforts to restore the species to the wild have proceeded in the 1990s and 2000s form the primary themes of this book. To set the stage for a consideration of these matters, we begin our presentation with a basic description of the species and a discussion of its ecological role as a giant scavenger.

Important Characteristics of the California Condor

The California Condor reigns today as the largest soaring bird of continental North America. Rarely flapping in flight, it is most famous for its magnificent appearance in the sky, where it substantially exceeds our native eagles, both in its size and in the grandeur of its aerial maneuvers. Although it is a bird without true vocalizations, the air passing through its finger-like primary feathers as it flies creates a steady hissing sound, audible from surprising distances. Often the first indication one has of an approaching Condor is an eerie crescendo of these wing sounds as a bird courses nearby above mountain ridges and meadows. Under favorable wind conditions, the Condor can exceed speeds of 40 mph in extended glides and cover nearly 150 miles in daily flight activities. Except around nests, roosts, and food sources, a Condor rarely lingers for long in any one location.

With a wing span often reaching nine to 10 feet and an average weight of nearly 20 pounds, the California Condor is indeed a monster among contemporary flying birds (pl. 2). Mostly black in coloration, it has long, triangular white feather patches on the undersides of the wings, short white bars on the topsides of the wings (pl. 3), and a largely naked head covered with baggy, wrinkled skin that is mostly bright orange in adults and dark gray in juveniles. At close range, a distinctly hooked tip is visible on the upper bill, and in adults, a bristly dark saddle of very short feathers crosses the forehead in front of the eyes (pl. 4). The feet are long and gray, and the heavy toes end in modest blunt claws, quite unlike the massive sharp talons of eagles and other birds of prey. Together these primary physical attributes characterize a species that does not closely resemble any other living bird, although certain of its characteristics are shared by other large soaring species.

The scientific name of the California Condor, *Gymnogyps californianus,* literally means naked vulture of California, referring to the general absence of feathers on the bird's head and neck and to its primary recent range in California and Baja California. The bird's present common name, California Condor, did not appear in early writings on the species and became widespread only in the mid-nineteenth century. Prior

Plate 2. In flight, adult California Condors are easily recognizable by distinctive long white feather triangles on the undersides of their wings.

Plate 3. Viewed from above, the wings of flying California Condors are marked with short white wing bars.

Plate 4. Air sacs of the head and neck region are commonly inflated in aggressive and sexual contexts.

to that time the species was most usually referred to as the California Vulture, the Royal Vulture, or simply the Vulture. "Condor" came into general use largely because of similarities of the species to the Andean Condor *(Vultur gryphus)* of South America.

Throughout its substantial former range, and continuing until the present, the California Condor's role in natural communities has presumably always been that of a highly social scavenger, feeding mainly on the flesh of dead mammals discovered from majestic soaring flight high above the ground. No records exist of the species capturing living prey in the wild, although one early record describes a Condor whose stomach was filled with the remains of mussels that may have been taken alive. In pursuing a scavenging lifestyle, the Condor has evolved a whole suite of adaptations that maximizes its abilities to find, compete for, and ingest carrion, and much of this first chapter is devoted to an examination of these specializations.

Taxonomically, the California Condor is a member of the family Vulturidae (or Cathartidae)—the New World vultures—a group believed to be closely allied with the storks in ancestry, but quite unrelated to the superficially similar Old World vultures, which are near relatives of hawks and eagles. The California Condor's closest living relative is the Andean Condor, which, although colored quite differently, is a near twin in size and habits (pl. 5). Other smaller members of this family include the King Vulture *(Sarcoramphus papa)* of Central and South America, and the Black and Turkey Vultures (*Coragyps atratus* and *Cathartes aura,* respectively), which occur throughout much of the Western Hemisphere.

Despite its impressive size, the California Condor falls considerably shy of being the largest known flying creature of all time. Certain flying reptiles, or pterosaurs, from the age of dinosaurs were much larger. For example, *Pteronodon longiceps,* which once coursed the inland seas of Kansas, had a wing span reaching 22 to 25 feet (pl. 6). Even more astonishing was

Plate 5. Although similar in size to California Condors, male Andean Condors are colored quite differently and possess a conspicuous head comb.

Quetzalcoatlus northropi of Texas, whose wing spread extended nearly 40 feet, roughly four times that of the California Condor. The weight of *Q. northropi* has been estimated to have reached nearly 200 pounds, roughly 10 times the bulk of a Condor. The fossilized remains of this incredible beast were first described only in 1975 by Douglas Lawson, then a student in paleontology at the University of Texas. *Quetzalcoatlus* and all other pterosaurs became extinct by the end of the Cretaceous, some 65 million years ago.

The California Condor is also considerably smaller than a number of extinct birds known as teratorns. The teratorns, close relatives of both condors and storks, were much more recent creatures than *Quetzalcoatlus,* with some member species still in existence about 10,000 years ago. The most impressive of the teratorns was *Argentavis magnificens,* a species

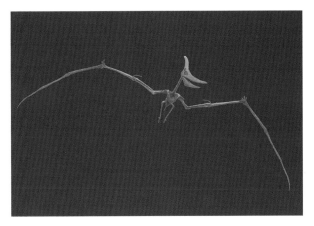

Plate 6. The largest of all flying creatures were flying reptiles, or ptero-saurs, from the age of dinosaurs. The skeleton of *Pteranodon longiceps* had a wing spread as large as 22 to 25 feet.

with an estimated wing span of about 23 feet and weight of about 160 to 170 pounds, the largest known flying bird of all time. *Argentavis* was a native of at least Argentina, but the full range it occupied is not known.

Other teratorns once occurred in North America. One of these— *Teratornis merriami*—had a wing spread of approximately 12 feet and is one of the more common creatures represented in the late Pleistocene tar deposits of Rancho La Brea in Los Angeles (pl. 7). Another— *Teratornis incrediblis*—described from fossils in Nevada and California, had a wing spread of about 18 feet and thus was only somewhat smaller than *Argentavis.* Evidently, the California Condor was far from alone in the skies during the Pleistocene and probably spent much of its time glancing backward and upward over its shoulders at some pretty awesome associates.

Thus, despite its renown as an avian giant, the California Condor barely qualifies for membership in the pantheon of the most impressive flying creatures of all time, and although it now reigns as the largest soaring bird of our continent, it has

Plate 7. As depicted at the Los Angeles County Museum, the extinct Merriam's Teratorn *(Teratornis merriami)* may have had a naked head, but this is not certain. Whether this species was primarily a scavenger or a predator is debated.

achieved this status only through the extinction of some considerably larger flying vertebrates. In fact, it appears that the contemporary form of the California Condor is slightly smaller than the form of the species extant in the late Pleistocene, a reduction that might reflect the disappearance of many of its larger competitors and food species since that time.

Were *Quetzalcoatlus northropi* and the teratorns also scavengers like the California Condor? The *Q. northropi* bones discovered included only one wing of a single individual, so it is presently impossible to judge whether the rest of this species' anatomy might have been consistent with scavenging. However, Douglas Lawson (1975) also found more complete bones of a smaller *Quetzalcoatlus* that has been generally referred to as *Quetzalcoatlus* sp. and that may have been either a separate species or, conceivably, the alternate sex of *Q. northropi*. This smaller creature had a long neck reminiscent of the neck of a modern-day griffon vulture, but much longer and not as flexible. Reasoning from this resemblance and from the fact that the *Quetzalcoatlus* bones were found far inland from the ocean and associated with the bones of huge sauropods, Lawson suggested that these pterosaurs may indeed have fed on the carcasses of other giant reptiles of their time. *Quetzalcoatlus* sp. had long bill-like jaws that apparently lacked teeth, but whether the tip of the upper bill was hooked, like that of typical scavenging birds, is unknown, as no specimen of a complete upper bill has yet been located. Some paleontologists think that the long bill of *Quetzalcoatlus* sp. was most likely straight at the tip, not hooked, and that it may have served to catch fish or to probe for living mollusks and arthropods, rather than for scavenging. *Quetzalcoatlus northropi* may have had a bill similar to that of *Quetzalcoatlus* sp., but especially if the two were different species, their bills and diets could have been quite dissimilar.

At least some of the teratorns were known to have hooked bills, but whether their bills were used primarily for taking carrion or living prey has likewise been a subject of debate. *Teratornis merriami,* for example, had a massive hooked bill and was very likely a flesh-eater, but from details of jaw structure, Kenneth Campbell and Eduardo Tonni (1981) have suggested it was more likely a predator than a scavenger. Nevertheless, many large predatory birds of today, for example, the Golden and Bald Eagles (*Aquila chrysaetos* and *Haliaeetus leucocephalus,* respectively) of North America readily take carrion as well as living prey, so even if *Teratornis* and *Argentavis* were mainly predators, it seems as likely as not that carrion formed a significant part of their diets. As pointed out by Peter Mundy and his associates (1992), the huge gape (mouth opening) of *Argentavis* and other teratorns might even have allowed these species to focus on eating a very special kind of carrion—sizable bones—today the primary diet of the extraordinary Bearded Vulture *(Gypaetus barbatus)*, also known as the Lammergeier, of Africa and Eurasia.

Regardless of what their diets were, the possibility that some or all of these early giants depended on carrion is generally consistent with what is known about the ecology and morphology of present-day flying scavengers. Many of our largest contemporary flying birds are scavengers, and as we describe below, there are a variety of reasons why aerial scavengers tend to evolve large size.

Scavenging as a Lifestyle

Achieving a basic understanding of scavenging is a necessary first step in gaining an appreciation of the role of the California Condor in natural communities. Accordingly, we believe it is important to have a general look at this lifestyle—examining the adaptations allowing species to become successful scavengers and the constraints faced by the creatures that adopt such a diet. Carcass-feeding is a subject that has been especially well studied by David Houston of Scotland and Peter Mundy and his associates of Africa; much of what we present on this subject derives from their perceptive research.

Vertebrate carcasses, despite their odious reputation, represent favorable food sources from a number of standpoints. When fresh, they are generally comparable to the best predator-killed prey in nutritional quality, and for species that are good competitors, they can offer jackpot quantities of food. And although carcasses that have suffered substantial decay may not always be the most savory of foods, they are at least food sources that cannot fight back or pose risks of debilitating injury. Nevertheless, carcasses do represent challenges of other sorts that need to be addressed by the creatures exploiting them.

For a number of compelling reasons, the scavenging lifestyle has probably never been especially common as an exclusive way of making a living. Vertebrate carcasses tend to be relatively scarce food sources, and few species specialize on them, presumably in part because of competition with already established specialized scavengers and with other opportunistically scavenging species, and in part because carrion is difficult to locate on a consistent basis unless a carrion-consumer has well-developed capacities for moving long distances quickly and with low energy expenditures. This requirement for efficient long-distance travel probably represents the core reason why specialized vertebrate scavengers of today are limited to flying creatures and why the only flying creatures that depend exclusively on carrion are large species capable of extended soaring flight. This is not to deny that many nonflying vertebrates and vertebrates of more modest flying abilities scavenge opportunistically—but these species characteristically take carrion as a sideline to other more reliable ways of procuring food.

In any event, within contemporary avian communities, the exclusive and near-exclusive scavengers, the vultures, are relatively few in numbers of species. In the Western Hemisphere there are presently just seven such species, compared to several thousand other bird species. In the Old World there are only 15 such species, again opposed to several thousand other bird species. Nevertheless, certain vultures do reach substantial overall abundance in numbers of individuals. In North America, for example, more than a million Turkey Vultures are commonly counted each year along migration routes south of the U.S. border. And in the Serengeti Plain of Central Africa, David Houston (1979, 1980) has documented that it is indeed the vultures that have become the major consumers of vertebrate flesh, substantially outranking carnivorous predator/scavengers such as Lions *(Panthera leo)* and Spotted Hyenas *(Crocuta crocuta),* both in numbers of individuals and in total food consumption.

A general scarcity in numbers of scavenger species may also have been the case among the pterosaurs of the age of dinosaurs—the aerial predecessors (but not ancestors) of birds. Assuming for the moment that *Quetzalcoatlus northropi* and *Quetzalcoatlus* sp. were truly specialized scavengers, they may have been virtually the only specialized scavengers among these ancient reptiles. Other smaller pterosaurs—and there were once a great many species in this group—are believed to have had very different diets. For example, *Pteronodon longiceps,* whose wings spanned 22 to 25 feet, and whose jaws were also toothless, is generally believed to have been an active predator of marine fishes, judging from fossil remains of whole fishes found within the rib cages of some specimens. Fish remains have also been found in association with other long-jawed pterosaurs. In contrast, many small pterosaurs had short and wide jaws well armed with sharp teeth. These species were likely also predacious carnivores, but exactly what prey they took is uncertain. Their victims may have ranged from insects and amphibians to early mammals and birds. One bizarre pterosaur with featherlike teeth on its lower jaw may even have been a plankton feeder. No pterosaurs have yet been found that likely fed on fruits, but some paleontologists speculate that such species may also have existed.

Thus, just as in bird communities of today, it appears likely that among pterosaurs the specialized scavenger lifestyle, if it existed at all, was at best a limited one, involving very few species. Among the more than 100 species of pterosaurs discovered, only the *Quetzalcoatlus* species have been suggested as reasonable candidates for scavengers. Further, if they were indeed scavengers, it seems likely that in pterosaurs, as in birds, specialized scavengers may have evolved mainly from active predatory forms.

Why predatory creatures might sometimes evolve into specialized scavengers, rather than remain predators that scavenge opportunistically, may lie primarily in aspects of food availability and the trade-offs in flight capacities between specialized scavengers and predators. Thus, as has been ably presented by David Houston (1979), the maximal long-distance soaring efficiency that is desirable for scavengers apparently can be achieved only by sacrificing the maneuverability in flight needed by active predators. Notably, it is characteristic of the large vultures that they lack finesse in landing and in executing rapid changes of flight direction at close quarters. They also find it relatively difficult to take off from the ground, generally requiring something of a running start to become airborne.

Nevertheless, when favorable conditions of carcass availability develop, natural selection may favor some species abandoning active predatory existence and concentrating on maximizing flight efficiency. The species best positioned to become specialized scavengers are quite naturally species that are already well adapted to feed on meat and already possess many of the adaptations needed for scavenging that we discuss below.

The scavenging habit as an exclusive lifestyle makes sense only under conditions where large vertebrate carcasses are at least moderately common and accessible. Thus if the *Quetzalcoatlus* species were truly scavengers, they probably could not have evolved before the great diversity of herbivorous dinosaurs had evolved in the Jurassic and Cretaceous. Likewise, the giant scavenging birds of terrestrial habitats today would have lacked an appropriate food supply until the evolution of large grazing mammals, which mostly followed the development of widespread grasslands around the world during the Oligocene, about 22 to 38 million years ago. Indeed, fossils indicate that Condors and other large vultures did not appear until about 15 million years ago in the Miocene, when such grazing mammals rose to prominence.

During the Pleistocene, which started about two million years ago and was the last geological epoch before the present epoch, there were many more species of large grazing mammals in existence than there are today, and correspondingly, there were considerably more species of probable scavenging birds. The close of the Pleistocene, however, brought massive extinctions of large mammals in the New World, and only a few of the large scavengers and predator/scavengers dependent on these mammals were able to avoid parallel extinctions. The end of the Pleistocene saw the disappearance of all the native teratorns and many of the New World vultures, although two of the larger species, the California Condor and the Andean Condor, managed to persist in especially favorable regions for foraging.

Many more scavengers were able to survive the end of the Pleistocene in the Old World, especially in Africa, which did not experience the massive mammalian die-offs seen in the New World. The large grazing mammals that still occur in Africa today represent a basically intact Pleistocene fauna, and not surprisingly, there are now more than twice as many vulture species remaining in the Old World as in the New World.

Relationship of Body Size and Bill Shape to Diet

The living vultures of today, both those of the New World and those of the Old World, are all medium-large to very-large species, although they are not the only large contemporary birds (consider swans, ostriches, and penguins). As already discussed, large size is a general advantage in pursuing a diet of carrion as it allows relatively rapid and wide-ranging soaring flight, facilitating the finding of food that is erratically distributed and fairly rare. Large size is also commonly an advantage in the rough and tumble competition for food that occurs among scavengers assembled at carcasses, and it can be a crucial advantage in giving scavengers the power to tear into the hides of carcasses, some of which are very difficult to penetrate. Finally, large size is generally correlated with relatively low metabolic rates, which are of value in avoiding starvation in species that cannot count on finding food regularly. The largest avian scavengers of today, among them the California

Condor, are known to sometimes survive more than a month without eating, although they need to average a full meal at least every several days to maintain a stable body weight. Thus, there are a variety of reasons why scavenging birds tend to be relatively large and why relatively large birds are often scavengers.

In addition to large size, all known specialized scavengers also have hooked bills. A hooked bill allows the ripping of small pieces off large food items, but like large body size, it is not a characteristic limited to carrion feeders. In fact, one extraordinary hook-billed vulture of today—the Palm-nut Vulture *(Gypohierax angolensis)*—has ventured into the vegetable world for a major part of its diet; another whole family of hook-billed birds—the parrots—feeds almost exclusively on vegetable material. Clearly, large hooked bills have evolved repeatedly among birds, and this anatomical feature alone, like many other anatomical features, does not give an unambiguous indication of diet. Unfortunately, the feeding habits of long extinct creatures such as pterosaurs and teratorns are often difficult to prove beyond all doubt simply on the basis of anatomy. Only when their gut contents are recognizably fossilized with them, does it sometimes become possible to know the feeding habits of extinct forms with near certainty.

With some of the largest birds still alive today—the condors of the New World and the large vultures of the Old World—direct observations indicate that carrion is one of the most important food types for birds of considerable size. The California Condor of today is an exclusive carrion feeder, so far as is known, although it is possible that it also once fed occasionally on defenseless seabird eggs and nestlings, as has been directly observed of its close relative, the Andean Condor of South America. The fossilized California Condor bones found together with bones of colonial seabirds on several of the Channel Islands off Santa Barbara and Ventura suggest this possibility.

Other Specific Adaptations for Scavenging

In addition to large size, soaring flight, and hooked bills, a number of other adaptations typify specialized scavenging birds. These include excellent eyesight, largely naked heads, long necks, large crops, feet with short claws adapted for walking and running, low reproductive rates, slow development of maturity, long lives, relatively great learning ability and intelligence, poorly developed vocalizations, resistance to bacterial toxins, feeding of young by regurgitation, equal division of labor between the sexes in reproduction, and nearly equal sizes of the sexes. These adaptations are found quite consistently in specialized avian scavengers regardless of their specific ancestors, and indeed, specialized scavengers appear to have evolved repeatedly and convergently from a variety of ancestral bird groups that share some of the above adaptations. The specialized scavenging birds we call vultures—both the New World and Old World vultures—represent only an extreme development of these characteristics. Less thorough tendencies to feed on carrion are found in many other contemporary families of carnivorous birds, and the scavenging adaptations listed above can be found in various stages of development in various species within these families.

A look at species exhibiting intermediate levels of adaptation to carrion feeding can help illustrate the selective factors involved. A good example is the Marabou Stork *(Leptoptilos crumeniferus)* of Africa (pl. 8), which commonly feeds at large mammalian carcasses, although it also takes many other sorts of food, including live amphibians and fish, which it normally swallows whole. Marabou Storks are very large soaring birds and, in line with their tendencies to take some carrion, exhibit a strong tendency toward naked heads and feet adapted for walking. They lack a distinctly hooked bill, however, which is an essentially universal trait in obligate scavengers and is also

Plate 8. Although the Marabou Stork often feeds on carrion, its long straight bill is poorly adapted for ripping meat from carcasses. It usually procures carrion by stealing meat chunks taken from carcasses by vultures, here mainly African White-backed Vultures *(Gyps africanus)*.

nearly universal in raptorial birds that do not swallow their prey whole. Marabou Storks have great difficulty in removing flesh from carcasses with their long straight bills but presumably have not evolved a hooked tip to the bill because it would be a hindrance in other feeding activities that are more typically storklike. Significantly, most of the carrion eaten by Marabou Storks is not flesh ripped directly from carcasses but carrion chunks stolen from other scavengers around carcasses, and for this purpose a straight bill serves perfectly well. Conceivably, the teratorns could likewise have taken carrion by piracy, even if their bills were best adapted for predation.

Another species with partial adaptations to scavenging is the Common Raven *(Corvus corax)*, a frequent associate of the California Condor at carcasses (pl. 9). The largest member of the family Corvidae, this species is highly attracted to vertebrate carrion and is often the first species to discover carrion sources. However, it also feeds on a great diversity of

Plate 9. Common Ravens, here assembling at a calf *(Bos taurus)* carcass in the San Joaquin Valley foothills, are often the first species to discover a carcass.

other food types, including the eggs of Condors on occasion. Although the tip of its upper bill is slightly hooked, which gives some aid in carrion feeding, the bill of this species is not fully adapted for this task, and ravens have considerable difficulty in detaching flesh from large carcasses. Ravens are capable of soaring flight, but they commonly use flapping flight to progress over the terrain, and although this may provide some constraints on the sizes of their foraging ranges, it also permits foraging under poor soaring conditions such as those often found in early morning, late afternoon, and in winter. Like the specialized scavengers, ravens can regurgitate food for their young, but they often carry food in their bills. Also like the specialized scavengers, their feet are primarily adapted for walking rather than seizing prey, and their intelligence is acknowledged to be especially well developed. Thus the Common Raven comes fairly close to exhibiting a full set of adaptations to scavenging, although it lacks the reduced feathering of the head, the long neck, and the strongly hooked bill of typical scavengers.

Plate 10. Adult Crested Caracaras feed on the carcass of a road-killed Armadillo *(Dasypus novemcinctus)* in Florida.

Still another species with partial adaptations to scavenging is the Crested Caracara *(Caracara plancus),* a member of the family Falconidae, a group of raptors composed largely of active predators (pl. 10). Crested Caracaras, in line with their tendencies to take some carrion, have hooked bills, feet fairly well adapted for walking, and partially naked heads. They also exhibit a relatively high degree of intelligence reminiscent of the intelligence of vultures, and possibly related to the complexity of their habits. Also like vultures, they are mostly silent, and what vocalizations they emit are largely given only in the immediate nest vicinity. Nevertheless, Crested Caracaras still find much of their food as active predators and have stopped short of evolving the very long necks typical of vultures. Moreover, they find much of their food from flapping flight rather than the soaring flight typical of the vultures. Also unlike the vultures, they typically transport food in their bill or talons and do not feed their young by regurgitation.

Thus Crested Caracaras, like Common Ravens and Marabou Storks, have only a partial set of adaptations for scavenging, and all three species also feed heavily on prey captured alive. These species differ greatly in appearance and in just which of their characteristics are best adapted for scavenging. These differences may relate in large part to the physical characteristics of their most recent ancestors and the nature of their alternative foods.

The specific advantages offered by the various characteristics listed above for scavengers are, for the most part, obvious and sensible. Excellent eyesight is surely helpful in finding food that is widely dispersed and sometimes difficult to see from the air. Naked heads greatly reduce the potential for feather fouling in species that commonly shove their heads into the slimy innards of carcasses. Long necks allow removal of food from far inside carcasses, and hooked bills greatly increase the rate at which scavengers can rip food from carcasses. Large crops also permit the rapid ingestion of large quantities of food, a crucial capacity in the highly competitive social environment of carcasses.

The advantages of feet adapted for walking and running are not quite as obvious, but because carcasses are almost always found on the ground, an ability to move with speed and agility on foot can greatly aid scavengers in competing with one another for such food. In contrast, many predatory birds that use their feet to kill and transport prey exhibit long curved talons, which are awkward for running. In virtually all vulture species, the extent of adaptation of the feet for walking and running is sufficiently strong that they are unknown to carry objects such as food or nesting material with their feet. Likewise, the teratorns had feet that were evidently adapted for walking, not for killing prey, and assuming they were predators, they may normally have used their bills to kill prey.

Low reproductive rates, slow achievement of maturity, and long lives (i.e., low mortality rates) are characteristics that are normally found together. They represent an emphasis on quality, rather than quantity, in evolution, and species possessing these characteristics generally exhibit quite stable, but often not very large, populations. Under favorable environmental conditions, such species often do very well at surviving through the ages, but they can become highly vulnerable to extinction if there are major changes in their environments. They often lack the reproductive potential for fast evolution of new characteristics to adapt them to new conditions, and even modest increases in mortality rates can place them at risk.

A substantial fraction of the endangered animal species of today are just this sort of species, and many are finding it difficult to cope with the numerous changes our own species has brought to the world. Such species often cannot tolerate mortality rates higher than about 10 percent per year because their reproductive rates are so low and they take so long to achieve maturity. The California Condor is one of the most vulnerable of all, as it normally does not breed until six to eight years old, raises only a single young in a breeding season, and often does not breed annually because its young have such a long period of dependency on their parents. Single-egg clutches and extremely long breeding cycles are also standard for other very large vulture species. The Bearded Vulture is the only large vulture to commonly lay clutches of two eggs, though it normally raises only a single chick per clutch.

The tendency for scavenging birds to possess relatively well developed learning abilities and substantial intelligence may well relate to the complexity of social interactions they face with other members of their own species and other species around carcasses. Scavengers rarely have food items to themselves for long and have to develop abilities to find and exploit foods that often involve evaluating the behavior of other scavengers in complex ways. The California Condor, for example, rarely finds food on its own but relies on watching other species, such as Turkey Vultures and Common Ravens, to discover carcasses. In addition, once Condors get to carcasses, they often have to deal with the intricacies of competing with other large birds, such as eagles, whose massive talons can pose severe threats.

How Condors respond to these competitors is variable and evidently depends on a diversity of complex factors, including whether or not other Condors are present at a carcass. In feeding, it can sometimes be adaptive for a Condor to collaborate with other members of its species, especially when it attempts to feed at a carcass whose hide is difficult for a single individual to penetrate. At other times it may be wiser for a Condor to exclude conspecifics, especially when dealing with small carcasses. The complexity of behaviors demanded of scavengers and the time it takes them to become skilled in finding and competing for food may be important factors in why they take so long to mature. Until they have several years of experience, individuals may simply be too inefficient in procuring food to make it worth their while to attempt reproduction.

Poorly developed vocalizations are another characteristic of the specialized scavengers, although reasons for this trait are unclear. The New World vultures even lack a fully developed syrinx (voice box) and are capable only of a variety of hisses and snorts audible at close range. Similarly, many of the Old World vultures are nearly mute, although some make primitive barking- or chittering-type vocalizations. Possibly the tendency toward minimal vocalizations in scavengers is correlated with a relatively greater reliance on visual communication signals in these generally open-country species, which often react to each other at distances too great for vocal signals to be of much use. Rudimentary vocalizations may also relate to the relatively nonterritorial behavior of vultures, as territorial advertisement is one of the most important functions of vocalizations in many birds. Vocal means of communication are most valuable for species, especially territorial species, that occupy densely vegetated habitats too cluttered for visual signals to travel very far. In this context, it is relevant to note that Peter Mundy and his associates (1992) report that the Palm-nut Vulture of Africa, the only forest-adapted vulture species in the Old World, is a relatively vocal vulture species.

Resistance to bacterial toxins is a trait one would predict in species that deal with carcasses that are also food for microbes of decay. The toxins excreted by such microbes are aggressive adaptations that help them compete for food with other scavengers. Although this is a field that has been minimally researched, it appears that at least Turkey Vultures are almost fully resistant to botulism, and it is hard to conceive how other scavengers might persist without developing similar abilities to handle this and other microbial threats.

Feeding of young by regurgitation from the crop may have a variety of advantages for scavengers. Considering the long distances that scavengers have to forage from nests, the moisture content and quantity of food they bring to their nests may be maximized by carrying it in their capacious crops rather than by carrying it in the bill or talons. This may be especially true because scavengers most often remove food from carcasses in small chunks that would be relatively difficult to assemble as large, manageable masses for transport in the bill or feet, especially in the highly competitive social environment that often exists around carcasses. Truly obligate scavengers, with their feet adapted for walking, are poorly equipped to carry food in their feet in any event, and birds carrying food in the bill or feet are especially vulnerable to piracy by other flesh-eating birds before they get back to their nests. Crop transport of food may also be aerodynamically superior to other means of transport, thereby facilitating long trips between feeding and nesting areas.

The division of labor between sexes in scavenging birds differs strikingly from that in raptorial birds. In many raptor species, females are considerably larger than males and commonly do essentially all the incubation of eggs and brooding and feeding of small young, whereas males specialize in foraging for their families. In scavenging birds such as the California

Condor, males and females are close to equal in size and share almost equally in incubation, foraging, and feeding duties.

Many hypotheses have been put forth as to why different raptor species show or do not show sexual size differences, but as yet no consensus on what adaptive forces may be most important has emerged. Our own view is that the food supplies faced by various species may often be the primary controlling factor in giving or not giving rise to advantages in divergence of sizes and roles of the sexes. Among various raptor species, the extent of sexual size differences is strongly related to the extent to which a species feeds on birds, and this may relate to an expansion of food supplies made possible by a sexual divergence in size in bird predators, but it is difficult to envision how sexual divergence in size might offer significant increases in food supplies for scavengers. Interestingly, although the overwhelming majority of raptor species exhibit a pattern of females being larger than males, this female bias is not consistent in scavengers. In fact, males are generally larger than females in both California and Andean Condors, whereas most other vultures in both the New World and Old World have females that are slightly larger than males.

Thus, in all important respects, the California Condor is a typical specialized scavenging bird. Its large size is a characteristic found in many other scavengers, and its largely naked head, relatively long neck, large crop, hooked bill, and feet adapted for running and walking are also expected characteristics, as are its low reproductive rate, soaring flight, lack of substantial sexual dimorphism, and lack of well-developed vocalizations. From the standpoint of adaptations, it is difficult to understand where the idea ever got started that the California Condor was a senescent species heading inexorably to extinction "with one foot, even one wing, already in the grave." As we shall see, the vulnerability of this species to extinction does not appear to stem from any lack of appropriate adaptations to its scavenging lifestyle, but from entirely extrinsic and novel mortality factors introduced into its environment by our own species—mortality factors that, through no fault of its own, the Condor could not reasonably be expected to have any adaptations to counter.

THE CALIFORNIA Condor, no less than any other species, is constrained by its nutritional needs. But because of the peculiarities of its diet, it faces especially challenging problems in obtaining food. As a scavenger of vertebrate remains, the Condor has to deal with a food supply that is often highly unpredictable in abundance and distribution. And as a flesh-eater, it must find ways to supplement its normal fare to assure adequate amounts of one nutrient, calcium, that is very scarce in meat and yet crucial for bone development and other functions. As we shall see in this chapter, historic Condors utilized enormous foraging ranges to satisfy their nutritional needs. Yet those ranges were largely separate and distant from their nesting ranges—a situation necessitating regular long-distance commutes. The inherent energetic inefficiency of such an arrangement makes good sense only in light of nonforaging factors that contribute to the bird's use of space. In this chapter we consider a variety of forces that have likely interacted to produce the movement pattern of the historic population.

Diet Diversity

From the paleontological research of Steve Emslie (1987), the historic research of Carl Koford (1953), and the more recent studies of food remains in Condor nests by Paul Collins and his associates (2000), the diversity of foods taken by California Condors is now quite well known. Recorded vertebrate foods for historic Condors have included carcasses of whales (various members of the order Cetacea), salmon *(Oncorhynchus* spp.), Cattle *(Bos taurus),* Sheep *(Ovis aries),* Mule Deer *(Odocoileus hemionus),* Tule Elk *(Cervus elaphes nannodes),* Grizzly Bears *(Ursus arctos horribilis),* Horses *(Equus caballus),* Mules *(Equus caballus* × *asinus),* Burros *(Equus asinus),* goats (*Capra* spp.), Domestic Dogs *(Canis familiaris),* Domestic Cats *(Felis catus),* Domestic Pigs *(Sus scrofa),* Coyotes *(Canis latrans),* Bobcats *(Lynx rufus),* Mountain Lions

(Felis concolor), rabbits *(Lepus californicus, Sylvilagus audubonii),* Striped Skunks *(Mephitis mephitis),* California Ground Squirrels *(Spermophilus beecheyi),* California Sea Lions *(Zalophus californianus),* kangaroo rats (*Dipodomys* spp.), Long-tailed Weasels *(Mustela frenata),* and Gray Foxes *(Urocyon cinereoargenteus).* In addition, studies of food remains in nests indicate that the species commonly consumed shells of marine mollusks and barnacles, presumably to help satisfy its needs for calcium. Some of these marine shells were likely gathered from ocean shores, but many may have been collected from subfossil deposits on inland hillsides, which are common in southern California.

The list of foods taken is long and varied but is most notable for its emphasis on mammalian species, with very few records of other vertebrates. Although no rigorous quantitative data exist on which species have been the most important foods, Carl Koford estimated that more than 95 percent of the diet might be represented by Cattle, Sheep, Mule Deer, Horses, and California Ground Squirrels (pl. 11). In a study of bones found in Condor nests, however, Paul Collins and his associates found that these five species accounted for only about 64 percent of the individuals identified, and that in addition to these species, the Condors were taking a great diversity of small- to medium-sized mammals, as well as occasional reptiles and birds.

Plate 11. A pair of California Condors approaches a calf carcass, the most commonly observed food of the species in recent years.

During the Pleistocene, the California Condor evidently fed on many huge mammals that have since disappeared from North America, including mammoths (*Mammuthus* spp.), camels (*Camelops* spp.), and native horses (*Equus* spp.). Direct documentation of the Pleistocene diet of the species has come from the bone remains found by Steve Emslie in Condor nests that date from this period (pl. 12). Located in cliff caves just above the Colorado River deep within the Grand Canyon of Arizona, these nests were identifiable as such from the remains of nestling Condors and fragments of Condor eggshell they contained. Like recent nests of the species in California, they also contained many mammalian bone fragments, presumably derived, at least in part, from carcasses that the Condors fed upon.

Emslie very reasonably attributed the prehistoric disappearance of the Condor from portions of its Pleistocene range east of the West Coast to the disappearance of much of its

Plate 12. The inner gorge of the Grand Canyon in Arizona provided nest caves for California Condors during the Pleistocene.

original food supply in this region. Although the causes of the disappearance of many of the giant creatures of the Pleistocene in North America are still under debate, the close coincidence in time between their disappearance and the arrival of humans in North America has led a number of researchers to suggest that humans were responsible.

Altogether, it does not appear that the Condor was ever a specialist on particular vertebrate species; probably it has always taken various kinds of vertebrate carrion mostly on an opportunistic basis, in accordance with availability, detectability, and penetrability. Aside from the advantages of feeding on carcasses that are relatively small and fresh (discussed below), Condors have probably had little to gain from being overly fussy in their food choices, as carrion tends to be a relatively scarce resource and most carcasses are roughly comparable in their nutritional characteristics.

Quantitative Food Needs

Studies of captives indicate that a California Condor maintaining its body weight normally eats a little more than a pound of meat each day and can hold approximately three pounds of meat in its crop. These figures indicate that an individual must ingest a full meal every two to three days to maintain body condition and on a daily basis must consume about 7 percent of its body weight in food. Similar rates of food consumption have been measured for other captive large vultures. Condors in the wild, of course, may show different food-consumption statistics, depending on the comparative energy expenditures of free-flying birds versus captives. Very likely the food consumption of wild Condors is somewhat higher than that of captives, although it has never been measured. Daily energy consumption of large, free-flying raptors whose flight includes a substantial amount of flapping is generally about twice that of captives of the same species. The difference may be much less for birds such as Condors, which rely on energy-efficient soaring flight. Thus Phil Kahl (1966) has estimated that a free-flying Marabou Stork *(Leptoptilos crumeniferus),* a soaring species, needs only 50 percent more food than the amount required by a captive of the same species. We estimate that free-flying Condors may need to average close to one full meal every two days to maintain body condition.

Finding and Competing for Food

Condors normally forage only during the midday hours, presumably because the winds and thermals necessary for extended soaring flight are often unavailable earlier and later in the day. At their overnight roosts on tall trees and cliffs, it is common to see the birds engage in leisurely preening and sunning activities until midmorning, before wind conditions become favorable enough for flight (pl. 13).

As the hours progress, the birds commonly begin to make abortive short forays into the air, apparently testing flight conditions, followed by immediate returns to perches. Once one bird does finally commit to extended flight, however, others quickly follow.

Carl Koford (1953), in his studies in the 1930s and 1940s, determined that daily flight periods of Condors generally lasted about seven to eight hours in summer and five to six hours in winter. Needs for food are especially great in winter, and opportunities for finding food are restricted because of fewer daylight hours available for foraging than at other times of year, frequent inclement weather, and generally lower temperatures. Not surprisingly, the species normally confines molt of its flight feathers to the period from spring through fall. This timing maximizes flight capacities during the winter months and ensures that the energetic demands of molt come during a relatively nonstressful time of year.

In locating food, California Condors apparently focus pri-

Plate 13. With the first rays of the morning sun, California Condors commonly extend their wings in a sunning posture, facing either toward or away from the sun.

marily on hunting in relatively open grassland areas where carcasses of large grazing mammals are most likely to be found and are most visible from the air. Foraging Condors normally circle and sail at a considerable altitude above the ground, often exploiting the updrafts along ridge tops and other topographic features and quickly taking advantage of thermal cells that develop erratically over the terrain. They almost never flap, except when taking off or landing, and because of their steadiness in flight often look more like distant airplanes than birds.

The Condor is rarely the first scavenger to discover a carcass, and the species relies mainly on observing the activities of other scavengers such as Common Ravens *(Corvus corax)* and Turkey Vultures *(Cathartes aura)* to find food. It is a large enough bird to displace most other scavengers when it arrives at a carcass, and when carcasses are sizable it can usually obtain a full meal even when it is a relatively late arriver. From the air, other scavengers gathering around a carcass are quite conspicuous, and when they or other Condors drop out of the sky to a carcass, they provide a visual signal of a feast that can be seen from many miles away.

Being first at a carcass is not necessarily the best strategy for a scavenger, as the first individuals arriving are faced with evaluating the safety of landing at the carcass when it may not be clear if the carcass is truly dead (and not simply a sleeping animal) and when it may be difficult to determine whether large and dangerous mammalian predator-scavengers are resting concealed in vegetation nearby. Later on, when many avian scavengers are working vigorously on a carcass, risks such as these are presumably much lower, as advertised by the presence of earlier-arriving scavengers already at work on the carcass.

First-arriving scavengers also often face a difficult task in penetrating the hides of many mammalian species, sometimes entailing considerable energy expenditure with little food to show for it. Those arriving later are often spared this

Plate 14. Turkey Vultures and other vultures of the genus *Catharres* are the only vultures known to find food by a sense of smell.

energetic investment, as the carcass by then has often been opened for exploitation. Thus, so long as a late-arriving species can compete well for food in the carcass environment, it may benefit by being somewhat slow and cautious in approaching a carcass. Only when the number of scavengers assembled is large enough to exceed the number of individual meals that can be obtained from a carcass does it become crucial to be one of the first to arrive.

No evidence exists that Condors can locate carcasses by smell, and it appears that such a capacity is most valuable for those scavengers that feed on very small dead animals and are faced with finding food in regions where it is difficult to see the ground from the air. Among the vultures, only species in the New World genus *Catharres*, most notably the Turkey Vulture (pl. 14), are known to be capable of finding food by odor, as has been most thoroughly investigated by David Houston (1986). This capacity, which even allows the *Catharres* vultures to locate food that has been buried underground, has made it possible for these species to forage successfully in dense forests. In turn, other scavengers, such as Black and King Vultures (*Coragyps atratus* and *Sarcoramphus papa*, respectively), successfully utilize the same habitat by watching, following, and parasitizing the foraging activities of the *Catharres* vultures.

No vultures with keen olfactory abilities are known among

Old World species, and it is presumably this difference that has led to a virtual absence of carrion-feeding vultures in the forests of that region. David Houston (1984a) has suggested that the difference between the New World and Old World scavenging communities in this respect may have been the evolutionary result of much better carrion supplies in New World forests, in part due to the abundance of sloths in tropical New World forests, but also due to much greater competition for carcass food supplies in the Old World from scavenging flies. Houston has observed that carcasses not discovered by vertebrate scavengers are commonly consumed by fly larvae within three days in the Old World, whereas they commonly last much longer in the New World.

Conceivably, a sensitive ability to find carcasses by odor first evolved in the genus *Cathartes,* and because this ability quickly led to the spread of the genus throughout the New World, other New World vultures may never have had much impetus to evolve the same capacity, being able to rely on the *Cathartes* vultures to discover food for them. However, although California Condors do follow Turkey Vultures to food in open areas and can presumably follow them to food in forested areas, there are relatively few records of them at carcasses in the latter habitats (Stoms et al. 1993), possibly because of the difficulty and inherent risk a bird with a greater than nine-foot wing spread faces in maneuvering through the tree branches. Alternatively, the near absence of records of Condors feeding in densely forested areas could be largely an observational artifact of the difficulty humans have in detecting Condors in such habitats.

Once on the ground at or near a carcass, California Condors rarely hesitate in displacing smaller species such as Common Ravens and Turkey Vultures from the feast; however, they usually show little inclination to challenge Golden Eagles *(Aquila chrysaetos).* A Golden Eagle can generally fill its crop in much less than an hour, and when one has gotten to a carcass ahead of a Condor, the Condor normally waits at a dis-

Plate 15. An immature California Condor (in background) and a Golden Eagle wait passively at a distance for another Golden Eagle to finish feeding on a calf carcass.

tance for the Eagle to finish (pl. 15). Unfortunately, because Golden Eagles usually feed only as single birds at a carcass, Condors can sometimes be faced with enduring long waits as successive Eagles arrive in the vicinity and take their turns feeding. Condors faced with substantial numbers of these birds waiting near a carcass sometimes give up without getting any food and return to the skies, possibly in an effort to find less well attended food sources elsewhere.

Like the Condors, Golden Eagles arriving at a carcass generally wait for an actively feeding Eagle to finish before attempting to gain control of a carcass. Incoming individuals of both species normally just perch on the ground a few yards distant from the actively feeding bird, apparently looking for signs that it is reaching satiation and losing interest in defending the carcass. Presumably for both Condors and late-coming Eagles alike, it is usually wiser to wait than risk damage from the massive talons of an Eagle in complete possession of a carcass. Eagles working on carcasses are extremely aggressive to challengers until they become sated. Thus even

Plate 16. Condors, especially dark-headed immatures, usually do not challenge Golden Eagles for food, but when they do, they risk attack and severe injury from the massive talons of the Eagles.

though Condors weigh about twice as much as Golden Eagles, they are usually subordinate to the Eagles in feeding. In contrast, Condors near their own nests are characteristically viciously aggressive to Eagles and do not normally tolerate Eagles in the vicinity, consistently driving them off in vigorous aerial chases.

The dominance of Eagles over Condors at carcasses is not absolute. On occasion, adult Condors do displace Eagles aggressively from carcasses, possibly when the former are especially hungry and the latter not so hungry, and it should not be forgotten that Condors themselves have impressive weapons in their razor-sharp bills, even if their feet are not equipped with eaglelike talons. Further, Condors are remarkably agile birds on their feet and capable of running with surprising speed, so a contest between the two species over food is not always destined to be won by the Eagle. Juvenile Condors, however, rarely challenge Eagles and rarely, if ever, are capable of winning such a contest (pl. 16). There are indeed records of juvenile Condors evidently killed by Eagles at carcass sites.

In within-species interactions, juvenile Condors are normally subordinate to subadults, and the latter are subordinate to full adults, in the "pecking order" that develops around carcasses. The establishment of clear dominance relationships among different-aged Condors may reduce the aggression that would otherwise occur and may well be facilitated by the substantial visual differences that exist among various age groups, particularly in head color and in the degree of whiteness of their under wing triangles (pl. 17). In threatening other Con-

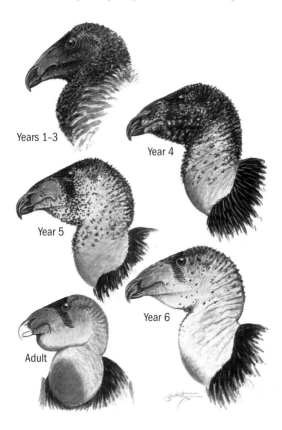

Years 1-3

Year 4

Year 5

Year 6

Adult

Plate 17. Head colors of California Condors change with age, as illustrated by John Schmitt. Full adult status is achieved at 6–8 years.

dors, a Condor typically displays its white under wing triangles with partially opened wings, and its head colors are likewise accentuated by the inflation of air sacs in the head region.

Behavior While Ingesting Food

When beginning work on a carcass, avian scavengers, including Condors, generally start with the most accessible tissues such as the eyes, tongue, and intestines reachable through the anal opening (pl. 18). However, if the carcass has already been opened up in other anatomical locations by other scavengers or has wounds in other locations, the birds do not hesitate to begin their efforts at these other points of entry. The intact hide of a large mammal is generally very difficult for them to penetrate, although a number of Condors working in concert at a weak point can sometimes tear through such a barrier. Such efforts are often concentrated in the belly region.

It is generally much easier for Condors and other scavenging birds to feed on small carcasses than on large carcasses because the former are not nearly as well protected by thick hides. Accordingly, foraging Condors often ignore conspicuous full-grown steer carcasses in an apparent search for carcasses that are easier to feed on. Although the carcasses fed on by Condors do include full-grown steers (perhaps mainly when other foods are unavailable), they feed much more commonly on smaller carcasses such as calves, Sheep, and California Ground Squirrels.

For the most part, Condors only occasionally find and feed on prey remains left by large natural predators such as Black Bears (Ursus americanus) and Mountain Lions. Instead their most commonly documented foods are animals dying of accidents, disease, or birth complications; stillborn mammalian infants; and the remains of game species (including gut piles) left by human hunters.

Condors also show an apparent preference for reasonably

fresh carcasses, as opposed to those badly decomposed, although this preference has not been studied in the detail that it has been in Turkey Vultures. In following the fates of Domestic Chicken *(Gallus domesticus)* carcasses of known stages of decay in the wild, David Houston (1986) found that Turkey Vultures in Panama normally did not find the carcasses until they were dead for at least a day, when they were first beginning to have a noticeable odor. Most carcasses were consumed while still relatively fresh (one to three days old), and the vultures commonly bypassed four-day-old carcasses, probably because of their advanced state of decay, not because the vultures failed to locate them.

Under favorable conditions a Condor can generally fill an empty crop with food in about 20 minutes, and from a distance it is often possible to detect how close a Condor may be to satiation by the degree of distension of its crop. Although it has been alleged that Condors can eat enough food to be unable to fly, we have never witnessed this nor obtained good

Plate 18. Black Vultures feed at the carcass of a Tapir *(Tapirus bairdii)* in Belize. The hides of large mammals are difficult for vultures to pierce and are usually first penetrated via the mouth, eyes, and anal opening.

evidence of it occurring under good flying conditions. Ian McMillan (1968) described once finding a fully fed Condor that apparently could not get aloft from the ground under still-air conditions, but the crucial factor was likely the still-air conditions, not the bird's full crop. We have seen Condors with apparently empty crops having similar difficulties under still-air conditions.

The structure of a Condor's tongue, which is equipped with rows of backward-pointing spines along its margins, aids in quickly removing flesh from carcasses. Meat is ripped free by the bill in relatively small chunks that are immediately swallowed as the bird works progressively into a carcass, often ultimately turning the hide inside out in the process.

Normally Condors limit their intake largely to soft tissues and ingest very little hair or hide. As a result they cast (regurgitate) pellets of indigestible hair infrequently and not on the daily basis seen in many other carnivorous birds. However, Condors also eagerly ingest small pieces of bone in an apparent effort to satisfy their needs for calcium, a nutrient that is not found in high concentrations in mammalian soft tissues. Unfortunately, most bones in large vertebrate carcasses are too large for ingestion by Condors, and the smaller bones, such as teeth, are not always easy to tear from their moorings. Consequently, Condors have a strong tendency to ingest small bone pieces wherever they run across them, not just at carcasses, and they sometimes visit the piles of bones left long after large carcasses have completely rotted away. Moreover, their apparent preference for small carcasses may in part be a reflection of the relatively greater amounts of ingestible bone found in them. This tendency for the birds to ingest hard bonelike materials in carcasses may predispose them to ingestion of lead ammunition fragments in hunter-shot carcasses—one of the most important threats to the species, as we discuss in later chapters.

A constant visual search for bone is very likely the root cause of the fact that mollusk shells and human artifacts, such as small pieces of light-colored plastic, glass, and metal, are sometimes found in Condor nests. This same phenomenon has been particularly well studied at nests of the Cape Vulture *(Gyps coprotheres)* of South Africa (pl. 19). Evidently both Condors and these large Old World vultures often confuse light-colored human artifacts with bone materials and are especially likely to ingest them when they are having trouble finding adequate amounts of real bone.

As researched intensively by Peter Mundy and John Ledger (1976, 1977), Cape Vultures normally obtain bone materials as fragments chewed up at carcass sites by Spotted Hyenas *(Crocuta crocuta)*. The extermination of hyenas across large portions of South Africa has had a major impact on the vultures of those regions by depriving the latter of their normal bone supplies. At the affected colonies of Cape Vultures, artifacts such as bottle caps and small pieces of glass and plastic have been abundant in nests, accounting for as much as 45

Plate 19. In the 1970s and early 1980s Cape Vultures in the Magaliesberg colonies of South Africa suffered severely from calcium deficiencies. Chicks in nests experienced frequent bone breakage and often failed to fledge successfully.

percent of the bone or bonelike materials found. Moreover, the nestlings at such colonies have exhibited major problems such as faulty bone development and impaction of their crops with trash, frequently leading to multiple fractures of their wing bones and a failure to fledge successfully (pl. 20). These problems have not developed at colonies where hyenas still persist in the vicinity, and the incidence of human artifacts in nests at these latter colonies has been much lower (generally averaging about 8 percent of bone or bonelike materials). Once they became aware of the calcium problems in hyena-free regions, Mundy and Ledger and their colleagues were able to correct the difficulty almost completely by establishing "bone restaurants" for the birds in the vicinity of the colonies, and chick bone development has since returned to normal in these colonies.

The incidence of human artifacts in wild Condor nests of the 1980s ran at a modest 12 percent of bone or bonelike materials, suggesting that Condors had not experienced calcium

Plate 20. This calcium-stressed Cape Vulture chick from the Magaliesberg colony of South Africa suffered multiple fractures of its wing bones, and its wings rehealed in a grotesque and non-functional pattern.

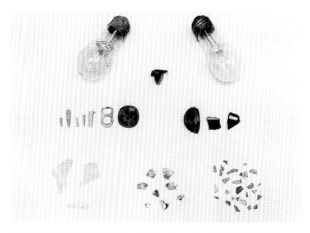

Plate 21. Human artifacts found in Condor nests during the 1980s included small pieces of metal, glass, and plastic presumably brought in by the birds, as well as flashbulbs from one site visited by photographers in 1941.

stress problems to the extreme extent documented by Mundy and Ledger for some colonies of Cape Vultures (pl. 21). No wild Condor chicks in the historic wild population of the 1960s through 1980s were observed to suffer bone breakage during development, so difficulties in obtaining calcium are unlikely to have been severe enough to have affected reproductive rates of the species significantly. Carl Koford (1953), however, documented one chick suffering bone breakage in one of its wings in 1939. Whether this instance was caused by calcium deficiency, by a mishap during banding of the chick, or by some other accident is not clear from existing records, but the substantial quantity of bone fragments found by Koford in this nest suggests that calcium deficiency was not the most likely explanation. Further, as described in chapter 9, two wild-reared Condor nestlings in the recent release program have had terminal problems with impaction of their digestive systems by fragments of glass, metal, and plastic.

Plate 22. One California Condor pair of the early 1980s commonly foraged on foot along the shores of an inland lake where they were strongly attracted to Styrofoam objects, often manipulating them in their bills.

One pair of Condors we watched repeatedly beachcombing along the shores of an inland lake in the early 1980s was often seen picking up white plastic objects such as Styrofoam cups and manipulating them in the bill (pl. 22). These birds may have been accustomed to searching for fish bones or other carrion along shore, and although we never saw them actually ingest a plastic object, they were obviously highly attracted to such objects and went far out of their way to inspect and pick them up. Their failure to ingest these items may have resulted from their detecting the softness and flexibility of Styrofoam objects with their bills.

At Condor nests under close observation in the 1980s we repeatedly watched nestling Condors sifting through nest-cave substrates with their bills and ingesting any light-colored objects they encountered, including small pieces of white excrement adhering to pebbles. Apparently, their ability to discriminate true bone from other hard objects somewhat resembling bone was imperfect, just as in adults.

Foraging Regions

In addition to foraging extensively along the shores of the Pacific Ocean, the historic population of California Condors foraged heavily in the foothills and to some extent on the floor of the vast flat-bottomed Central Valley (including both the San Joaquin and Sacramento Valleys) that today comprises the major agricultural breadbasket of California (pl. 23). The foothills of the valley, too steep for agriculture, have traditionally been used for grazing Cattle and Sheep, and to this day their vegetative cover has remained largely in grasslands or grasslands mixed with oaks. Prior to the invasion of California by Europeans, these hills, like the bottom of the Central Valley, were the home of large herds of Tule Elk, Mule

Plate 23. The major twentieth-century foraging region for Condors was the foothills of the San Joaquin Valley. Used mainly for grazing livestock, these foothills formerly hosted strong populations of Pronghorn, Tule Elk, and Mule Deer.

Deer, and Pronghorn *(Antilocapra americana)*, and they still held good populations of Mule Deer in the twentieth century. Evidently, the Condors of the region were able to shift their diet from the original native ungulates to livestock as the former were largely killed off by early settlers.

Meanwhile, regular foraging along the shores of the Pacific Ocean largely disappeared from the Condor population by the early twentieth century, although causes for this decline are unsure. Even until the late twentieth century, Condors continued to nest within relatively easy flight distance of the ocean and at least some marine carrion has continued to be available along the ocean shores to the present, so the recent absence of beach foraging calls for an explanation. Possibly, the disappearance of beach foraging was primarily a result of strongly diminished quantities of beach carrion, as many marine mammals of California (for example Northern Elephant Seals *(Mirounga angustirostris)* underwent major human-caused declines in the late nineteenth century (Schoenherr 1992). Alternatively, the amount of disturbance of coastal habitats by a steadily increasing human population and the vulnerability of beach-foraging Condors to shooting might have been primary causes. Still another possibility is that beach-foraging Condors may have suffered especially severely from tendencies of the species to ingest dangerous human artifacts, such as sharp glass and plastic objects. Unfortunately, the loss of coastal foraging by the species occurred long before the first careful studies of the species were undertaken, so we can only speculate on what specific factors may have been most important.

By the mid-twentieth century, when the first major studies of Condors were conducted, the Condor population exhibited a pattern of movements that mainly involved commutes from nesting regions in the mountainous Coast Ranges, Transverse Ranges, and southern Sierra Nevadas to foraging regions in the San Joaquin Valley foothills (map 3). These commutes commonly entailed flights of 20 to 30 miles just to get to the

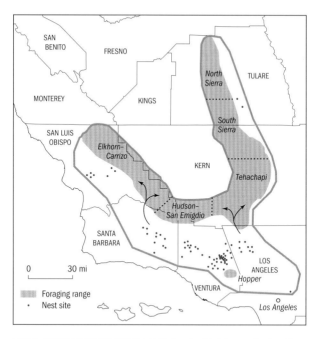

Map 3. Historic California Condor nest sites (red dots) were geographically separate from the Elkhorn–Carrizo, Hudson–San Emigdio, Tehachapi, South Sierra, and North Sierra foraging zones in the San Joaquin Valley foothills. The map illustrates the full range of the Condor in the 1980s (circled area). Arrows indicate common routes of travel from two nest areas to the main foraging grounds.

edge of the foraging grounds and often took the birds close to an hour to make. Moreover, even though apparently suitable cliffs for nesting also can be found in certain specific regions of the San Joaquin Valley foothills themselves, these cliffs evidently were not used by the population for nesting, either in historic or recent times (pl. 24). Because these cliffs were much closer to the primary food supply than the cliffs actually used for nesting, the failure of the birds to use them is another distribution anomaly that demands explanation.

Perhaps the best hypothesis as to why the Condors have

Plate 24. Possibly because of nearby dense populations of Golden Eagles, Condors are not known to have nested in otherwise suitable cliffs within the major foraging regions.

nested so far away from their main food supply is that potential Condor nests within the foraging regions may simply have been too vulnerable to predation by Golden Eagles, which are abundant in these regions. As we shall see in the next chapter, adult Condors characteristically leave their nests unguarded for most of the daylight hours during the latter portions of the breeding cycle, and their nestlings tend to wander on foot outside their nest caves from an early age, making them highly vulnerable to any Eagles that might see them. Actual predation attempts of Eagles on Condor nestlings have been witnessed, although they have been rare, probably because the known nests of Condors have been characteristically located in areas with very low Eagle populations. The tendency of Condors to nest far from the main foraging regions could well represent an adaptation to minimize predation on chicks, in spite of the energetic penalties involved in long commutes to the foraging regions.

Daily Movements of Breeders and Nonbreeders

While they are breeding, Condors are necessarily tied to fixed nest sites, and as was determined by intensive radio-telemetry and photo-documentation studies of the 1980s, foraging movements of historical breeding pairs generally involved commutes to the nearest portions of the main foraging range, searches up and down the foraging range from the points of entry, and later returns to their nests by a reverse set of movements. This pattern of movements led to strong emphasis of nearby portions of the foraging range in their foraging activities, as can be appreciated in map 3, which illustrates nests and foraging movements of selected historic pairs. Core foraging ranges for individual breeding pairs of the 1980s, as defined by areas including 95 percent of foraging-range observations for the pairs, averaged about 1,000 square miles, and pairs mostly foraged within 30 to 40 miles of their active nests, although they occasionally ranged as far as 90 to 110 miles from nests. These foraging ranges are much larger than those known for large predatory birds (which generally run from about 10 to 75 square miles) and instead are comparable to foraging ranges documented for large Old World vultures.

When not breeding, historic pairs of Condors still commonly frequented their nesting territories, but not as regularly as when breeding. As temporary nonbreeders, they wandered quite freely up and down the full foraging range known for the species, a region of about 2,700 square miles in the 1980s, and they often roosted far from their nests, giving relatively little emphasis to parts of the foraging range close to their nests.

Both when nesting and when not nesting, pair members commonly foraged together, except during parts of the breeding season when incubation and brooding duties demanded steady attendance of one adult or the other at the nest.

Unpaired Condors were not constrained by having to make frequent visits to nests, and they wandered fluidly through all portions of the foraging range at all times of year, roosting at night at traditional sites scattered throughout the range. Nevertheless, when not foraging, these nonbreeders (often young birds) occasionally visited breeding areas of the species. Presumably they gradually learned the full range of the species, including all nest sites, by such a process. Although recent fledglings confined their movements to the nesting areas from which they originated and to nearby portions of the foraging range, by the time the birds became subadults several years later, they were visiting essentially all parts of the species' range, a region including about 7,700 square miles in the 1980s.

At no time of the year were breeding pairs seen attempting to defend feeding territories from other Condors, and surely such an effort would have been totally impractical in view of the immense sizes of their home ranges. Instead, all birds, including breeders, were decidedly, though intermittently, social in their foraging movements and activities.

Movements Related to Seasonal Changes in Food Availability

Superimposed on the constraints on movements imposed by breeding, historic wild California Condors exhibited traditional seasonal shifts in emphasis of different parts of the range for foraging. For example, there was a clear historical emphasis on foraging in the southeast (Tehachapi) foothills of Kern County during the fall that was correlated with the deer-hunting season in that region (map 3). Similarly, extensive summer use of southwestern (Hudson–San Emigdio) portions of Kern County was correlated with the calving season for Cattle in that region—especially aborted calves resulting from epizootic bovine abortion (foothill disease). Primary use of the Hopper region above Fillmore was correlated with a winter-spring calving season. Thus, although some Condors could be found foraging most anywhere in the foraging range of the species at any season, the Condor population as a whole tended to focus its main foraging efforts in different regions at different times of year. The seasonal patterns in use of the foraging range appeared to be correlated primarily with geographic shifts in food abundance that repeated year after year.

Much more dramatic seasonal shifts in food availability have been documented for Old World griffon vultures in the Serengeti ecosystem of central Africa. Here, as studied by David Houston (1974a, 1974b, 1976), the main carcass supply is provided by enormous migratory herds of Wildebeest (*Connochaetes taurinus*), and the vultures feeding on the carcasses provided by these herds are faced with keeping up with the substantial movements of the herds with the seasons. That they are able to do so far more successfully than the territorial predatory mammals and crocodiles that also prey on these herds is apparently the key reason why the vultures have become the dominant carnivores of the region, both in numbers and total food consumption.

Similar great migratory herds of Bison *(Bison bison)* were known historically in the Great Plains region of North America, but there is no evidence, at least in historical times, that Condors ever took advantage of them the way the griffon vultures of Africa have exploited Wildebeest. The failure of Condors to develop such capacities may simply have been a result of the considerably longer distances traveled by the Bison herds than by the Wildebeest herds—distances potentially too great to be practical as regular commutes from nests for the Condors (for historic routes of Bison migrations, see Hornaday 1889). Thus, even though the great Bison herds did survive the Pleistocene extinctions, they may have represented too unreliable a year-round food supply for nesting Condors to have allowed persistence of Condors in the Great Plains region into historical times. Actual historical records of Condors feeding on Bison are virtually nonexistent, although bones found in Pleistocene Condor nests in the Grand Canyon of Arizona indicate that some Bison were eaten by the Condors of that time. During the Pleistocene, many additional (and now extinct) species of mammals also roamed the countryside and potentially provided alternative foods for the Condors at times of the year that Bison were too distant to be a practical food supply.

The seasonal shifts seen in Condor foraging range in studies of the twentieth century were not completely regular. In particular, Ian and Eben McMillan (in Miller et al. 1965) documented an apparently new pattern of use of one portion of the foraging range (northeastern San Luis Obispo and southeastern Monterey County) that developed in the 1940s. Prior to that time Condors were not commonly seen in this region, but they became regular visitors for a period of some years at that time, apparently correlated with substantial increases in carrion supplies.

Changes in Food Availability in Historical Times

By the nineteenth century, the food supply of the California Condor on the West Coast had become largely restricted to domestic livestock, with the near extermination of many species of native mammals, although the birds still continued to take some Mule Deer carcasses and sea mammals along the coast. In that era, much of California was used to raise Cattle for the tallow and leather industries, with the meat of carcasses largely left unutilized and available for scavengers. Because of the extent of the slaughter, this may well have been a time of major food abundance for Condors, although it was also likely a time when lead contamination of carcasses became a major stress on the population.

Through the twentieth century, however, the importance of livestock in the economy of California progressively declined, as agriculture came to dominate in the San Joaquin Valley, and grazing lands were progressively lost to farming and urbanization in many regions. These trends have continued to the present, and it is no surprise that many observers over the years have suspected that declines in food supplies might have been the primary cause of the declining population of Condors. Both Carl Koford (1953) in the 1930s and 1940s and Sanford Wilbur (1978b) in the 1970s documented continuing substantial declines in livestock food supplies, although Alden Miller and the McMillan brothers (Miller et al. 1965) argued in the 1960s that recent resurgences in Mule Deer populations might have largely compensated for declines in livestock. Through direct estimates of numbers of available carcasses through the seasons, these latter authors came to the conclusion that food supplies were unlikely to be an important limiting factor for the Condor population of the 1960s and that the Condor decline was instead due to excessive mortality factors that were not linked to starvation or food stress. Ultimately, direct evidence gained on reproduction, mortality rates, and causes of mortality in the 1980s was to largely vindicate the general view of Miller and the McMillans, as we shall see in chapter 6.

AS SEEMS appropriate for such a large bird, reproduction is a ponderous affair for the California Condor. Condors do not normally breed until six years old and then produce only one chick at a time, with each full breeding cycle lasting more than a year. Their normally low rate of reproduction is of conservation concern because it permits only very slow population increases at best and can yield stable or increasing populations only if mortality rates of the species remain very low. By comparison, breeding pairs of smaller vultures, such as the Turkey Vulture *(Cathartes aura)*, often produce two or three young per year. Because of their greater reproductive potentials, these species can achieve population stability or even increase in the face of considerably higher mortality.

In this chapter we focus on the basic reproductive habits of the California Condor, detailing the behavioral characteristics of historic breeding pairs, major events of the breeding cycle, and the characteristics of the nest sites chosen by the species. Chapter 6 presents a more quantitative treatment of reproduction, analyzing breeding effort and success as part of a consideration of causes of the historical decline. Condor reproduction was especially well studied by Carl Koford in the 1930s and 1940s (Koford 1953), by Fred Sibley in the late 1960s (Sibley 1968, 1969, and unpublished), and by diverse participants in the intensive research program of the 1980s (Snyder and Snyder 2000). The presentation here is based largely on the combined results of these efforts.

The opportunity to study Condor nests was surely one of the most exciting aspects of the Condor program of the 1980s, as the sites chosen by the birds for nesting were among the most spectacular and remote scenic locations in the region, despite their close proximity to Los Angeles and other coastal cities. Here it was possible to forget about civilization and to begin to see the world as Condors have no doubt seen it throughout their history as a wild species. Observations were made from blinds and lookouts that had unobstructed views of both the nests and their surroundings and that were positioned far enough away from nests (usually at least a quarter mile) to allow the birds the freedom to pursue their daily activities without interference. Observers camped out among Black Bears *(Ursus americanus)* and Mountain Lions *(Felis concolor)* and made dawn-to-dark observations, usually for periods of several days, before they were relieved by replacement observers in an effort to maintain constant daylight coverage of active nests. Two people in alternation were generally needed to keep each nest continuously monitored, and only by continuous coverage could rates of crucial activities be measured accurately and could rare but important events, such as interactions with potential predators, be reliably observed.

Nest observations demanded long periods of concentration when little was happening and necessitated careful attention to details. Male and female adults were very similar in appearance and only differentiated by subtle differences in feather patterns, especially differences in molt and feather damage, and by their roles in courtship and copulation behavior. Normally, there was ample time to identify which sex was in attendance at a site, but when both adults were present together in the nest vicinity and were moving about continuously, it was often a scramble to take accurate notes on what took place.

Events at nests were often unpredictable and ranged from nervous encounters between Condors and bears to sudden arrivals of military jet aircraft screaming low overhead in simulated combat maneuvers. Winter storms sometimes buried high-elevation nest sites, Condors, and observers in snow, and there were recurrent problems with dense fog and rain. Despite these impediments, the opportunity to study breeding Condors in detail yielded unique and valuable data obtainable in no other way and experiences that no one involved in these efforts will ever forget.

Courtship and Pair Formation

Condors do not normally become fully adult in feather and skin colors until they attain six to eight years of age, and at least in the wild they do not normally breed until this coloration is attained. Once formed, pairs are exceedingly stable in the species and normally endure until the death of one or the other adult; however, a Condor whose mate dies generally finds a new mate within a year or two and experiences only a temporary lapse in breeding. Condors rarely survive long enough to reach an age of declining reproductive capacities, although a terminal cessation in breeding was seen in one captive female (a bird of unknown age). On surgical examination, the reproductive tract of this female was found to have suffered physical degeneration, although she continued to show some reproductive behaviors with her mate after this point. As a minimum this bird bred for 16 years, but she may well have bred for many additional years before she was first identified as an individual.

Pair formation has been observed primarily during the fall and winter seasons and has involved several conspicuous behavior patterns: pair flights, mutual preening, and wing-out mating displays. In pair flights, birds in incipient or established pairs fly close together in coordinated soaring displays in which the two birds cruise in parallel through their prospective or established nesting territory (pl. 25). The birds fly slowly, adjusting constantly to maintain their tight spacing, and seem to make a great effort to remain side by side. From a distance, the two birds often seem to fuse together into a single larger flying object, and the enhanced size of this composite creature may serve to advertise the presence of a pair to other Condors to a greater distance than would be the case if they flew far apart. Although Condor pairs do not commonly engage in aggressive territorial defense, evidence from the 1980s suggests that they do space themselves out within the

Plate 25. In pair flights, adults fly in parallel close together and often appear to merge into a single flying object when seen from a distance.

broad nesting regions, so it may be important for pairs to be able to assess the locations of neighboring pairs efficiently.

Pair flights are often one of the earliest signs of pairing, but they do not always lead to the establishment of viable pairs. For example, during the spring of 1982, we watched a bird with very distinctive feather damage in its wings appear successively in four different established territories where it engaged in pair flights with members of three of the resident pairs. None of these apparent courtship attempts led to the bird's acceptance by any resident, and he (or she) soon gave up and left each territory.

In Condor pairs that progress beyond initial courtship, it is common to observe mutual grooming activities (pl. 26). In these, the birds perch side by side on a rock or limb of a tree and use their bills to nibble at each other's skin and feathers, especially around the head and neck—anatomical regions they cannot reach effectively with their bills in self-preening activities. Mutual grooming activities normally appear gentle, but we occasionally saw instances where a bird grabbed

Plate 26. Mutual preening in a pair of adults is usually focused in the head and neck region, where self-preening is difficult.

loose folds of a partner's neck skin in its bill and did not let go, even when its partner moved to new position (pl. 27). The bill of a Condor is an extremely sharp tool for severing meat and skin, so mutual grooming can presumably pose some risks if not pursued with judicious restraint.

Wing-out sexual displays are stately and colorful ceremonies that are usually performed on relatively flat surfaces, such as the ground or on cliff ledges, but sometimes take place on tree limbs (pl. 28). They are normally performed only by males, and a displaying bird extends its wings in partially drooped position, stretches its head and neck forward and downward, and slowly struts its feet up and down in exaggerated fashion. The displaying bird often sways back and forth and commonly circles its mate on foot when there is enough space for this. During display, which usually lasts about a minute, the air sacs in the head and neck region are often, but not always, inflated, revealing brilliant orange and pink patches of skin.

Plate 27a (above) and 27b (below). In mutual preening a Condor sometimes grips the loose neck skin of its mate tenaciously, failing to let go even when continued preening is exceedingly awkward.

Plate 28. A male California Condor gives a wing-out courtship display to his mate atop a giant sequoia in 1984.

Sexual display commonly leads to copulation, and in mounting, the displaying male steps onto the back of his partner, where he stands awkwardly with his wings flapping and with his bill grabbing his mate's neck skin, apparently to aid in balance. Mounting episodes, like wing-out displays, normally continue for about a minute, and when they are successful, they conclude with the male bending his tail around and under that of the female from one side to achieve cloacal contact. A female seeking to reject a copulation attempt simply moves away or nips at the male's head with her bill when he attempts mounting.

Nest Sites

Most nests documented for the California Condor have been caves or potholes in cliffs (pl. 29), although some sites have been little more than overhung ledges or crevices in boulder piles on steep slopes, and there have been two known nests that were cavities high above the ground in the trunks of giant sequoias *(Sequoiadendron giganteum).* Judging from their charred interiors, the latter had probably been produced by fires igniting large dead branches and burning into their attachments to the trunks. Evidently the Condor, like other New World vultures, prefers natural cavity sites offering some protection from the rain, relatively easy access from the air, flat substrates for eggs, space for at least two adult-sized birds, and ceilings high enough to allow the birds to stand up over their eggs. Aside from these general characteristics, Condors seem to have fairly broad tolerances in the kinds of sites chosen. No Condor nest sites have yet been found in dense thickets on flat ground, as has been documented for Black Vultures *(Coragyps atratus)* and Turkey Vultures, although it is possible that Condors nesting in Florida during the Pleistocene may have used such sites in the absence of any steep topography in this region. One early Condor nest in California (the first one ever documented) was evidently a hollow at the base of a large tree on a steep slope.

Condors are not known to bring nesting material into their nest caves, although both sexes do a certain amount of rearranging of litter near the egg location with the bill, gener-

Plate 29. Most Condor nest sites have been used repeatedly, but irregularly, through the years. The single egg is normally laid as far back in the cave as is feasible.

ally resulting in the formation of an ill-defined disk of leaves, sticks, gravel chunks, acorns, and woodrat (*Neotoma* spp.) droppings in the egg location. These disks of debris may function in limiting uncontrolled rolling of eggs, facilitating gas exchange, and absorbing some of the mechanical shock that eggs receive during the laying process.

Most nests studied closely have been used repeatedly by the birds over the years, although most often in irregular alternation with other sites, as the birds, at least in recent years, have usually changed sites in successive attempts, regardless of whether they have failed or succeeded in the sites. This tendency to change nest sites in successive nesting efforts may represent an adaptation to reduce parasite buildups in nests, and indeed some Condor nests studied have had heavy parasite infestations, principally ticks, lice, and bloodsucking bugs. Alternatively, frequent switches of nest sites may reduce risks of predation on nests. An active Condor nest develops a powerful and persistent odor over time, detectable by the human nose from many yards away. Odor-oriented nest predators, such as bears, may have less chance of locating eggs and nestlings if active sites are allowed to cool off between uses.

With respect to visually oriented nest predators, such as Common Ravens *(Corvus corax)*, the safest Condor nests appear to be deep and dark sites in which eggs are not visible from the entrances. Such nests are not always favorable in other respects, however. Some particularly deep and dark sites studied in the 1980s held clouds of gnats (*Dasyhelea* sp.) that were apparently highly irritating to incubating adults, causing endless head shaking.

Nest Investigations

Alternate nest sites of a pair are often distributed among a number of canyons and are often several miles apart. Prior to laying in a specific site, pairs generally check out many of their alternative sites over a period of several weeks to months. During investigations the birds enter the sites, move substrate around with their bills, and perch, copulate, preen, and sometimes roost in the vicinity. In a minority of cases pairs have quickly focused their attentions on single sites, so that it was apparent that these sites would likely become the sites used for egg laying. In most cases, however, pairs have continued to investigate a variety of sites up almost to the point of egg laying. During early and mid-stages of the nest investigation period it was usually very difficult to predict which site would actually receive the egg.

In a typical nest investigation, a pair spends up to several hours in and around a site before flying off to forage or to investigate a different site. So long as they stay together in movements, it is safe to assume that egg laying is not imminent. In the final day or two before egg laying, however, the male typically goes off to forage alone, leaving his mate in the nesting territory. At this stage the female commonly fills her crop at a nearby water source, and just before egg laying she often remains in her prospective nest site overnight for the first time. Nevertheless, observations of the 1980s indicated that a female sometimes continued to check a variety of sites even during this stage. When her mate eventually returned from the foraging grounds, he sometimes checked many of the various nest sites he and his mate had been investigating, in what appeared to be an effort to relocate his mate. This suggested that he was no better than we were in predicting which site the female might choose for egg laying.

Active nests are also visited by nonbreeding birds on an occasional basis throughout the breeding cycle. Although such visits are usually limited to perching in close proximity of the sites, visitors sometimes even land at the actual entrance of a nest, commonly without evoking strong aggression from the nest-site owners. Because most active nests received such visits from nonbreeders, it appears likely that most nest sites have been known to many birds in the population. Thus, it should not be surprising that long traditions of use have developed at many sites, presumably often involving successions of birds through the years, although specific sites are often unoccupied for many years between uses.

On one memorable occasion in 1984 we watched from a helicopter above the forest as three Condors flushed from the branches next to a well-concealed and inactive nest cavity in a giant sequoia—a site unused since 1950, so far as was known. Evidently the site was still familiar to at least some birds in the population, despite being fallow for many years and being almost invisible from the air. In 1968 Ian McMillan expressed fears that this site might have been permanently deserted by Condors because of lumbering operations and other forms of human disturbance in the area. But whether use of the site had actually been affected by human disturbance is unclear. The site's existence was obviously still known to the Condor population in 1984, and the characteristics of the site in that year evidently did not preclude its close examination by several birds. These apparently included one member of a pair of birds nesting in the same year in another sequoia tree only about five miles distant and both members of a pair nesting in a cliff site more than 100 miles distant in the same year.

Egg Laying and Incubation

As witnessed several times in the wild and in captivity, a female Condor typically lays an egg from a standing position and the egg drops to the substrate with some apparent force, although there are no cases known of an egg being damaged in the laying process. Characteristically, during egg laying the female faces away from the nest entrance with a hunched-over posture, and a series of tremors shakes her body immediately before egg expulsion. Observations in captivity indicate that she also emits a wheezing-squealing noise at the very moment the egg emerges. In the cases observed, incubation of the egg began almost immediately (within about five to 10 minutes) after laying (pl. 30).

Each breeding attempt involves only a single greenish

Plate 30. An incubating Condor rests the egg on its feet and normally changes position every hour or two. White excrement coats the walls of sites that have hosted Condor chicks in earlier years.

white egg that soon fades to pure white. Incubation normally lasts 56 to 58 days. If the chick hatches successfully it receives about five-and-a-half months of parental care before fledging, and after fledging it remains fully dependent on its parents for several more months before it begins to forage on its own. Thus, a full breeding cycle from the day of egg laying to the point of at least partial independence of a fledgling takes more than a year, and adults do not normally start a new breeding cycle until that point is reached. If an egg is laid early in one year (late January through mid-March) and the chick fledges successfully, the pair may start another breeding attempt late in the egg-laying season of the following year (late March through early May). Available data suggest, however, that a pair normally does not follow a late (March to May) egg laying with a breeding effort the following year unless the late-season breeding attempt fails.

In summary, a consistently successful wild pair can be expected to make breeding attempts in only two years out of three. An unsuccessful pair recycles more frequently, normally laying each year. Further, a pair that fails early in the breeding season frequently lays a replacement egg about a month later within the same breeding season—a process known as double-clutching or replacement-clutching. In some instances where a pair has failed early enough with a second egg, it has even laid three eggs (triple-clutching) within a single breeding season. Successful fledgings from second or third eggs, however, are unlikely to be followed by breeding attempts the following year.

Male and female adults of a pair share almost equally in incubation and brooding duties and feed their chick with essentially equal frequency. Pairs vary considerably in the frequency with which adults take turns at the nest, however. For four pairs we watched on a continuous basis in the incubation period in the 1980s the average lengths of incubation shifts varied from 2.26 to 5.89 days among pairs, and in one pair incubation shifts occasionally lasted as long as nine or 10 days. While one adult incubates, the other normally leaves the nesting area to forage, and while the latter is away, there are usually no signs of Condor activity in the nest vicinity. Thus finding nests during the incubation period is especially difficult, compared to other stages of the breeding cycle when adult activities around nests are much more conspicuous and frequent.

Hatching and the Nestling Period

The hatching process from first pipping of the eggshell to final emergence of a chick takes nearly three days in Condors, and during this period the adult in attendance shows keen interest in the egg, sometimes nibbling at the pip hole with its bill and apparently aiding the chick in its struggles to break free of the shell. This parental assistance may be important in helping the chick conserve its energy reserves through the long hatching process.

Chicks hatch with a wispy covering of white down and receive their first feeding soon after hatching (pl. 31). They are brooded almost constantly during the first two weeks, but thereafter brooding declines rapidly and virtually ceases during the daylight hours at about one month. By this time chicks are well along in developing a thick second coat of dark gray down and are apparently able to regulate their daytime body temperatures quite well without parental assistance (pl. 32). Nevertheless, parental attendance at the nest continues at night for about another two to three weeks.

Interestingly, the average rate of changeovers of adults at the nest increases dramatically at the time of hatching to near-daily frequency. For example, at one nest of 1983, adults changed places about once every 3.9 days during the incubation period but exchanged duties about once every 1.2 days during the first four weeks of the nestling period. As during incubation, the off-duty adult leaves the nesting area to forage while its mate attends the nest. The immediate dramatic increases in changeover rates seen after hatching at all nests suggest that the more relaxed changeovers during incubation may not have been dictated by such factors as food availability.

As brooding declines late in the first month of the nestling period, the overall attendance of adults at nests shows a rapid decline and then stabilizes at a much lower level from about the start of the third month through the rest of the nestling period. During the 1980s, nest caves were normally attended by adults less than 5 percent of the daylight hours in the late stages, and adults were present in the nest vicinity only about 25 percent of the daylight hours, leaving their nestlings unprotected from predators most of the time (fig. 1).

Feedings are frequent (averaging about once every two hours) during early stages, but this frequency drops steeply during the first few weeks. Data from nests of the 1980s indicated an average of about one feeding every 10 daylight hours

Plate 31. Newly hatched California Condor chicks have head colors ranging from yellowish to grayish and have bodies covered with wispy white down. (Photograph taken in 1941.)

Plate 32. At nearly two months of age a Condor chick has replaced its original white body down with a dark gray down and has begun to develop short dark down on the head. (Photograph taken in 1906.)

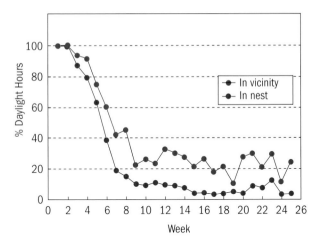

Figure 1. Based on data from five intensively studied Condor pairs of the 1980s, adults tend their nests almost constantly for the first few weeks of the nestling period. Nest attendance then declines to a much lower level at about two months through the rest of the breeding cycle. The top curve gives the average percent of daylight hours one or both adults are in the nest vicinity; the bottom curve gives the percent of daylight hours one or both adults are in the nest site itself.

after about two months, a rate that remained quite stable thereafter until fledging. Each feeding typically consists of several bouts of contact of the adult with the chick, during which food is transferred bill to bill by regurgitation. During the latter stages, chicks have been given no feedings at all on about one-quarter to one-third of the days of observation. Occasionally they were not fed for two consecutive days, although this occurred only 13 times in 605 days of observation of the five most intensively studied nests of the 1980s. Only one instance was recorded of a chick receiving no food for three straight days.

Overall, the feeding rates seen at various nests of the 1980s were quite uniform, and this uniformity did not in itself suggest any major problems in finding food for breeding. Never-theless, possibly because of differences in pair quality, or perhaps foraging range quality, or even differences in the crop capacities of various adults, there were some minor differences in consistency of feeding among the intensively studied pairs. The most consistent pair studied failed to feed its chick on only 15 percent of the days in weeks 10 through 25 of the nestling period and had no consecutive no-feeding days. The most inconsistent pair failed to feed its chick on 30 percent of the days in weeks 10 through 25 and had seven pairs of consecutive no-feeding days. Nevertheless, the female chick of this latter pair fledged successfully, and when she died a year later of cyanide poisoning from a coyote trap, her weight was excellent (8.4 kilograms) for a female.

Male and female adults feed their chicks with roughly equal frequency, with 542 male feedings and 531 female feedings tabulated overall at the intensively studied nests of the 1980s. Thus, just as with the near equality of their contributions to incubation, it appears that the two sexes invest almost exactly equally in chick care. In fact, the only striking difference in contributions of the sexes prior to fledging appears to be the production of eggs only by females. But because an egg constitutes only about 3 percent of the body weight of a female, this difference appears nearly negligible in the overall economy of reproduction through to fledging.

In addition to feeding their chick, the adult Condors also engage in exceedingly complex preening interactions with the chick. In these, both adult and chick twine their necks together and rub their heads and necks over each other's bodies, sometimes interspersed with bouts of aggressive nipping behavior on the part of the parent (pl. 33). An adult seeking to restrain a frantically begging chick often places one foot on the latter's neck, clamping it to the substrate. Foot-clamping of a chick's neck is also a common way for an adult to remove a nestling's bill from inside its throat at the end of a feeding.

During the long hours adults are absent from the nest vicinity, nestling Condors mainly divide their time among spells of sitting alert, preening their feathers, and sleeping.

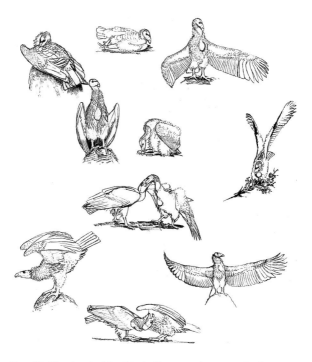

Plate 33. Sketches by John Schmitt illustrate a late-stage Condor nestling preening, sunning, exercising, playing, and interacting with a parent.

Older chicks also spend lesser amounts of time picking up, manipulating, and sometimes ingesting objects such as sticks, feathers, stones, bones, and leaves. They sometimes dig into the substrate with their bills and eat light-colored objects, apparently selectively, when they encounter them, possibly in an effort to satisfy calcium needs. Older chicks also engage in periods of exercise—flapping their wings, leaping about, rapidly turning while jumping, jabbing at their wrists with their bills, capturing objects in the bill and then whirling around and carrying them away, and so on. They also exhibit mock prey-capture behavior with their feet, stabbing out to stomp an object against the substrate without actually gripping it. The presence of this behavior in chicks, but not adults, suggests that the ancestors of Condors may have had at least partially predatory habits.

Fledging

Fledging is a gradual process in California Condors, and the time of fledging can only be defined in an arbitrary fashion, as chicks often begin to make walking excursions from their nest caves long before they can fly (sometimes as early as six weeks of age) and often leave the ground briefly during wing exercise long before they can manage a truly coordinated flight (pl. 34). Nevertheless, we believe it is useful to consider fledging to be the first flight that is ambitious enough to take a young bird beyond a walking commute back to the nest en-

Plate 34. At nest sites that permit easy movements on foot outside, Condor chicks begin to wander about on adjacent slopes as early as six weeks of age. Late-stage chicks have patches of white down on the topsides of their wings that flash conspicuously when they flap their wings in exercise and in begging for food. (Photograph taken in 1983.)

trance, necessitating an aerial return to the cave. As so defined, fledging flights were observed at three different nests of the 1980s and took place when chicks were five to six months of age (specifically 178 days, approximately 163 days, and an unknown age, respectively). A much earlier fledging that was documented occurring at about 145 days by Carl Koford (unpublished field notes) in 1939 apparently involved a youngster that fell accidentally from its nest cave, because the bird's departure from the nest cave (which was not directly witnessed) was not followed by any true flights for about another month (pl. 35). The first scrambling flight up onto the cliff near its nest by this bird was at 174 days of age, and perhaps this might better be considered the age of fledging.

Fledging flights directly witnessed in the 1980s all occurred in the absence of adults and covered distances ranging

Plate 35. At a low Condor nest of 1939 (arrow), Carl Koford documented a chick fledging prematurely at about 145 days of age, apparently falling accidentally from the site. Though it survived the fall, the chick took no more flights for another month.

from about 20 yards to 300 yards. These flights were exceedingly clumsy, with desperate flapping and dangling, backpedaling feet, followed by highly uncontrolled landings on whatever substrate first loomed in front of the youngster. This clumsiness disappeared only very gradually and, for example, was still conspicuous in one fledgling observed closely in March 1980—likely a bird that had taken its first flights at least four months earlier.

Fledging flights were seen as early as mid-September and as late as early November in the 1980s, although we suspected that a fledgling first found in 1982 that was exceptionally delayed in feather molt may have fledged as late as December 1981 (conceivably even January 1982) and could well have been the result of a second or third egg in 1981.

Following fledging of their youngster, parent birds continue to return periodically to the nesting territory both to feed the fledgling and to circle about. Fledglings often fly with their parents as they get older and show increasing tendencies to follow their parents back toward the foraging grounds when the latter depart from the nesting territory. In the early stages of fledgling dependency, however, such following flights always break off, with the fledgling returning to the nest vicinity. When parental visits are infrequent, fledglings often increase the altitude and range of their movements in the nest vicinity as the time since their last feeding increases.

A young female observed fledging on September 22, 1982, was first observed leaving her natal territory to accompany her parents all the way to the foraging grounds on February 1, 1983, more than four months after fledging. Her parents initiated a new nesting attempt with egg laying on approximately March 31, 1983. In the weeks immediately preceding egg laying, the adult female ceased feeding her daughter of the previous year and became conspicuously aggressive to her, repeatedly driving her from the nest vicinity. The adult male, on the other hand, continued feeding the fledgling and showed little aggression to her during the same period. Thus the weaning of this youngster evidently was initiated first by the female, a situation that may not necessarily apply to pairs that do not reinitiate breeding in consecutive years.

Natural Enemies
of Breeding Condors

As one of the largest birds in its range, the California Condor does not face a long list of enemies capable of overpowering it as an adult. Nevertheless, eggs and nestling Condors have faced significant predation threats from two other bird species, Common Ravens and Golden Eagles *(Aquila chrysaetos)*, and from several large terrestrial mammals, most notably the Black Bear in recent times. During the 1980s most nest failures occurred at the egg stage and were caused by Common Ravens, although Golden Eagles were twice seen in failed attempts to take nestlings. Both of the latter attempts were foiled by effective defensive efforts of adult Condors. Also during the same decade, a Black Bear was once watched as it apparently smelled a Condor nestling and attempted unsuccessfully to reach it, first by scaling the vertical nest cliff itself and then by climbing a tree in front of the cliff.

The principal defense of the species against bears (presumably including the Grizzly Bears *[Ursus arctos horribilis]* formerly resident in California) is no doubt the positioning of most nests on cliffs that are difficult for large terrestrial predators to climb. Of 72 recent and historical Condor nests that were assessed in the 1980s, 48 (67 percent) were in sites that we rated as completely inaccessible to large terrestrial predators, eight (11 percent) were in sites that we rated as possibly accessible to such predators with great effort, and 16 (23 percent) were rated as easily accessible to such predators. The surprise comes from the relatively frequent use of easily accessible sites when there was no apparent scarcity of potential

Plate 36. Below one nest cave in a high precipitous cliff of 1966 (see arrow), Fred Sibley discovered a chick that had evidently fallen to its death. This same nest cliff was the very last nesting location used by the historic population in 1986.

nest sites giving much better protection from terrestrial predators. Perhaps the relatively common use of accessible sites is a reflection of avoidance of the risks inherent in more inaccessible sites of chicks accidentally falling to their death during development. As mentioned earlier, a chick likely fell accidentally from the nest most intensively studied by Carl Koford in 1939, although it survived the fall, likely because the nest cave was not extremely high on the cliff. In contrast, Fred Sibley (unpublished field notes) found a freshly dead chick under a nest high in an inaccessible cliff of 1966, and it appeared likely that the chick had fallen to its death (pl. 36). In 1983, we witnessed many near falls of a nestling from another high and inaccessible site, and the chick was taken into captivity early, in part because of the evident risks.

In part, Condors address the risks from Golden Eagles by aggressively driving intruding individuals away from their nest vicinities (pl. 37). When they happen to be present near

Plate 37. Condors drive a Golden Eagle from their nest vicinity with a vigorous flapping chase.

Plate 38. Condors often drive Common Ravens from the nest vicinity, but their defenses against Ravens are less consistent, and usually far less effective, than their defenses against Golden Eagles.

Plate 39. Much more agile than Condors, Common Ravens commonly return to harass nesting Condors immediately after being chased off.

their nests, Condors are normally successful in driving Eagles away with frantic flapping chases. But during the latter stages of the nesting cycle adult Condors are normally not present in the vicinity of their nests for 75 percent of the daylight hours, leaving their chicks completely exposed to Eagle predation most of the time. The principal defense against Eagles appears to be the choice of nesting sites in regions where Eagles are rarely seen, especially regions far from the grassland foraging habitats normally utilized by the Eagles.

Attempts by adult Condors to actively defend their nests from Common Ravens have appeared less effective, and Ravens have commonly returned quickly to harass Condors after the Condors have driven them away (pls. 38, 39). Although Condors incubating eggs are normally successful in preventing Ravens from gaining access to their eggs, their attentiveness to their eggs has been less consistent than has been observed in other large vultures threatened by large corvids. In particular, Condors often leave their eggs unguarded for

Plate 40. Prairie Falcons sometimes nest close to Condor nests and can be helpful to the Condors by excluding other predators such as Golden Eagles from the vicinity.

many minutes during nest exchanges, during which both adults often circle over their nesting canyon. Reasons for this laxity are not obvious, and our best rationalization of this situation is that Condors may not have had enough time to evolve fully satisfactory defenses against Ravens because Ravens may formerly have been too rare to constitute a major threat and may have become a substantial threat only in recent decades. Quite good evidence exists that Ravens have increased appreciably in southern California in the past century, likely because of many new sources of food provided by ranching and farming activities and by road-killed vertebrates (Knight et al. 1993).

With respect to both Eagle and Raven predation, it appears that the Condors sometimes receive significant, although inadvertent, protection provided by another species nesting close by—the Prairie Falcon *(Falco mexicanus)*. This species aggressively excludes other large birds from the vicinity of its own nests, more or less indiscriminately (pl. 40). And although it is difficult for the Condors to enter their own nests when the Falcons nest close by (because of direct harassment by the Falcons), there has been no evident interest of the Falcons in preying upon the contents of the Condor nests, and when the Falcons have been present, the Condors have evidently been much less threatened by Eagles and Ravens. For example, essentially the only nesting of the CC Condor pair in the 1980s that was not characterized by frequent and intense battles with Ravens was one in which Prairie Falcons nested very close to the Condors. That the Condors might choose to nest near the Falcons has not been obvious, however, and may not normally be possible because Condors usually begin nesting long before the season Falcons choose their own nest sites.

JUDGING FROM the progressive shrinkage of its range, the California Condor experienced a continuing major population decline throughout the nineteenth and most of the twentieth centuries. No quantitative estimates of the size of the historic Condor population were offered until the twentieth century, however. Further, through most of the twentieth century, the population estimates that were put forth bore only speculative relationships to actual population sizes, making calculations of rates of decline in various periods a highly questionable endeavor. Little was known through much of this period of the patterns of movements of the birds, and the huge size and ruggedness of much of the species' range precluded the sort of thorough coverage that was needed to develop good population estimates by direct counting methods. Only in the 1980s did it finally become possible to accurately track the size of the population by a newly developed photographic method that at last resolved the major censusing difficulties. Unfortunately, by that time the population had declined to a mere remnant, confined to the region just north of Los Angeles.

Early Records

The earliest contact of humans with California Condors presumably occurred roughly 12,000 years ago, when our species first came to North America from eastern Asia. From archeological evidence in middens and caves, it is clear that Condors were symbolically important to early human cultures, and they still figured prominently in sacrificial ceremonies of Native American societies in early historical times, as discussed by Bancroft (1882) and Harris (1941). To a lesser extent Condors were also featured in cave art (pl. 41) and in the fabrication of ritual garments by Native American societies (pl. 42).

The earliest written accounts of the California Condor mostly trace to the end of the eighteenth and beginning of the nineteenth centuries when California was first explored and

Plate 41. A California Condor pictograph dominates the entrance of Condor Cave, a Chumash winter solstice observatory in Santa Barbara County.

Plate 42. Edward Davis of Mesa Grande displays a Native American ceremonial skirt of Condor feathers made from a bird shot in 1899.

settled by northern Europeans, although Harris (1941) documented a few earlier written descriptions of Condors by early Spanish explorers. The species was first described scientifically in 1797, based on a specimen, now in the British Museum, that had been collected by Archibald Menzies in 1792 (Grinnell 1932). During the early to mid-nineteenth century the Condor was often considered a locally regular or common species, sometimes occurring in gatherings of more than 100 individuals (Taylor 1859b, Cooper 1870, Harris 1941).

Nevertheless, by the first half of the nineteenth century it was already clear that the species was disappearing from some regions. For example, the Condors seen along the Columbia River bordering Oregon and Washington by the Lewis and Clark expedition of 1805 to 1806 were apparently gone just a few decades later (Cooper and Suckley 1859). Retrospective analyses by Koford (1953) similarly indicated that the northern California population was also in serious decline at an early date, followed by major declines in central California, southernmost California, and Baja California.

That the wild Condor population as a whole was declining seriously was only first clearly articulated in 1890 by James G. Cooper, the founder of the Cooper Ornithological Society. In an article entitled "A Doomed Bird," Cooper stated, "I can testify myself that from my first observation of it in California in 1855, I have seen fewer every year when I have been in localities the most suitable for them. There can be little doubt that unless protected our great vulture is doomed to rapid extinction." Although Cooper offered no numerical population estimates for the species, he was obviously convinced of its progressive disappearance, and he attributed the decline primarily to three causes: (1) poisoning resulting from predator-control efforts, (2) a decline in food supplies—Cattle and Sheep (*Bos taurus* and *Ovis aries,* respectively)—resulting from changing land use patterns, and (3) shooting. These themes were affirmed in many subsequent accounts of the plight of the species, though often with little indication as to whether they rested mainly on Cooper's assessment or on other information. Just how important these three stresses may really have been and how our understanding of the causes of the species' decline has changed over the years are the primary subjects of the next two chapters.

Following Cooper's account, there were a scattering of additional accounts around the turn of the twentieth century that lamented the decline of the Condor but gave no population estimates (for example, Bendire 1892, Beebe 1906, Hornaday 1911). Of these, several papers on the species published between 1906 and 1926 by William Finley were surely the most important and influential. Together with Herman Bohlman, Finley had carried out a periodic study of a nesting pair of Condors in Eaton Canyon, just north of Los Angeles, in 1906, and the photographs of the birds taken by these workers, currently archived by the Museum of Vertebrate Zoology in Berkeley, have become historic classics (pls. 43–45). Finley, like Cooper, was convinced of the continuing decline of the species, and like Bendire and Cooper, he attributed the decline mainly to poisoning and shooting. Although he made no speculations on the numbers of Condors still in existence, he assembled much of the existing historic information on the species in his accounts and added greatly to our understanding of the bird's reproductive habits from his own personal experiences.

First Population Estimates

The first numerical population estimates for the species were those by William Leon Dawson (1923), Joseph Grinnell (in Phillips 1926), and Harry Harris (in Wetmore 1933). Dawson, who was based at the Natural History Museum in Santa Barbara, had only very limited personal experience with the species in the wild, and his estimate of fewer than 100 birds and perhaps as few as 40 birds rests on an unknown and prob-

Plate 43. The most famous photograph of California Condors ever taken was a historic portrait by William Finley and Herman Bohlman of a pair close to its nest in Eaton Canyon, near Los Angeles, in 1906.

Plate 44. William Finley with the Eaton Canyon Condor chick and one adult in 1906. The adults were extremely approachable and attentive to their chick.

Plate 45. The chick taken from the Eaton Canyon nest suns near Finley's home along the Willamette River in Oregon. This bird was later transferred to the New York Zoological Park where it lived for eight years.

ably nonrigorous foundation. Grinnell's estimate of 150 birds (75 pairs) for the 1920s was based on some personal field experience with Condors but again on unknown methodology. Harris's estimate of 10 birds was likely nothing more than a wild guess, perhaps extrapolating from Dawson, as Dawson had predicted the effective demise of the species within about a decade of his estimate. As we shall see, all these early estimates were likely major underestimates of actual numbers.

Only with the studies of Cyril Robinson (1939, 1940) and Carl Koford (1953) were the first attempts at comprehensive counts begun. Robinson (pl. 46) was the deputy supervisor of the Los Padres National Forest and had a keen interest in the species. He organized efforts to collate sightings by Forest Service personnel and estimated 60 Condors for 1934 and 55 to 60 Condors for 1938 to 1939, based largely on summations of high counts for two regions—the Sisquoc and Sespe regions in the Coast and Transverse Ranges (Santa Barbara and Ventura Counties). Although these were traditional concentration areas for nesting and roosting of the species and were

Plate 46. Cyril Robinson was the deputy supervisor of the Los Padres National Forest in the 1930s and was the first person to attempt systematic counts of Condors.

both ultimately to become official Condor sanctuaries, they fell far short of including all important nesting and roosting regions of the species in recent times. Sporadic high counts limited to these regions likely missed many Condors, both within the regions and elsewhere.

Carl Koford's similar estimate of 60 Condors, apparently for the 1940s through 1950, was based in large part on his more extensive field experience in the same two regions. With financial support from the National Audubon Society, Koford (pl. 47) spent four years studying Condors for his Ph.D. at Berkeley, and he produced the first major monograph on the species. Koford was impressed that the largest single-flock sightings he documented (approximately 40 birds) were no larger than the highest simultaneous counts he achieved with cooperators in several locations, and he evidently concluded these totals were likely close to the total numbers of living Condors and added 50 percent to the high counts to arrive at an estimate of 60 birds. Nevertheless, his simultaneous counts were few and geographically limited and could easily have missed many Condors.

At about the same time as Koford's study (1942), two California Department of Fish and Game employees (D. D. McLean and J. S. Hunter) reported a single-day total of 122 Condors seen and counted successively at carcasses in the Antelope Valley and the Carrizo Plain. These observers felt, with some justification, that there was likely little overlap in the birds seen only a couple hours apart in the two regions, and thus their count suggested a considerably larger population than estimated by Robinson and Koford. Probably based mainly on this sighting, the California Department of Fish and Game roughly estimated a population of 150 birds for the same period. Carl Koford presented no analysis or discussion of the McLean and Hunter sightings and simply noted that their count in the Antelope Valley exceeded his total population estimate. As we shall see, another independent estimate of about 150 birds for 1940 to 1950 can be generated by backward extrapolations of data from the 1960s through 1980s.

Plate 47. Carl Koford examines a Condor nestling from his most intensively studied pair of 1939. Like Cyril Robinson he estimated 60 birds
for the wild population.

The next important population estimate for the Condor
was one of 40 to 42 birds for the early 1960s that was put forth
by Alden Miller (pl. 48), together with Ian and Eben McMillan,
in 1965. In a study sponsored by the National Audubon Society
and the National Geographic Society as a follow-up to Carl Koford's study, these workers mainly compared flock sightings of
the early 1960s to those of the 1940s, and because the large
flocks they documented in the 1960s averaged about 30 percent
smaller than those of the 1940s, they concluded that the total
population was also about 30 percent smaller than Koford's estimate of 60 birds for the 1940s—or roughly 40 to 42 birds.

Their estimate, however, was strongly dependent on the
accuracy of Koford's estimate, and it did not credit an apparently reliable count of 63 Condors seen on one occasion in
1961 by Bert Snedden, a rancher in the San Joaquin Valley
foothills who was intimately familiar with the species. Moreover, the estimate of 40 to 42 birds was soon superseded by
counts greater than 50 birds on the annual October Survey, a
massive simultaneous count of major portions of the Condor
range that was initiated by diverse interests in 1965 (Mallette

Plate 48. Alden Miller, here on a 1948 collecting expedition in Mexico, supervised Carl Koford's graduate research on Condors in the 1930s and 1940s and later supervised a follow-up Condor study conducted by the McMillan brothers in the early 1960s.

and Borneman 1966) and continued annually until 1981. Originally designed to employ standardized observing methods from the same locations over the years, the October Survey in fact utilized variable and generally declining numbers of locations and observers over the years, raising strong criticisms of how well it might measure trends (Verner 1978, Wilbur 1978b). In addition, there were some chronic difficulties in determining whether the birds seen from different observation points during the counts were the same or different birds, necessitating consensus analysis to arrive at probable total numbers of birds seen.

Nevertheless, as proposed by Fred Sibley (pl. 49), the first U.S. Fish and Wildlife Service (USFWS) researcher on the Condor (who conducted studies from 1966 through 1969), the October Survey did provide a reasonably secure minimum estimate of about 60 birds for the late 1960s, an estimate that is supported by backward extrapolations of the firm counts of the 1980s through comparisons of flock sizes in the late 1960s and 1980s. Moreover, in a general way the continuing October Survey left little doubt that the population was

Plate 49. Fred Sibley was the first Condor researcher for the U.S. Fish and Wildlife Service and conducted studies from 1966 to 1969. Sibley believed that 60 Condors still remained in the late 1960s, based primarily on the annual October Survey that was started in 1965.

Plate 50. Sanford Wilbur was the second Condor researcher for the U.S. Fish and Wildlife Service and conducted studies from 1969 to 1979. Wilbur believed that only about 30 Condors remained in 1978.

still declining strongly through the 1970s. By 1978, Sanford Wilbur (pl. 50), Sibley's successor in the USFWS Condor program, was estimating only about 30 birds left in the wild, an estimate that subsequent data have also suggested was probably very close to the actual population size.

Counts of the 1980s
and a General Assessment

Truly accurate counts of the wild population were not secured until a greatly expanded research program was initiated in 1980 involving participants from many diverse agencies and organizations. By 1982, a new method of counting Condors was developed through comparisons of large numbers of photographs that showed primary feather patterns of flying birds. This method made it possible to achieve continuous reliable individual identifications and counts of all Condors, regardless of where they were encountered in the range (Snyder and Johnson 1985). Counts from late 1982, when 21 birds were documented, until the last wild individual was trapped and taken into captivity in 1987, may well have been free from error. How these counts were obtained and analyzed are discussed in some detail in chapter 6.

By comparing flock sizes seen in various regions over the years it is possible to extrapolate backward from the firm counts of the 1980s to probable numbers present in earlier years. This process, as detailed in Snyder and Snyder (1989), gave strong support to Sibley's estimate of 60 birds for the late 1960s but produced much higher estimates for the early 1960s and 1940s than those given by Miller and his colleagues (1965) and Koford (1953). In particular, analyses of comparative flock sizes, region by region, suggested that the population of the 1940s may have been about 150 birds, as was indeed estimated by the California Department of Fish and Game at the time, and was later estimated as well by Fred Sibley. The population in the early 1960s likely included on the order of 100 birds.

In fig. 2 we present the major Condor population estimates from 1950 through the 1980s. No matter which estimates one may favor—those of the Koford-Miller-McMillan tradition, those of the October Survey, or those of the

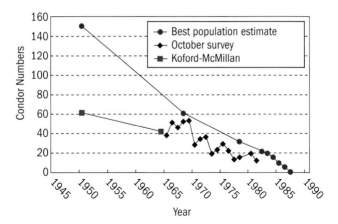

Figure 2. The principal Condor population estimates and counts from 1950 through the 1980s all indicated population decline. The estimates of Koford and the McMillans were much lower than those of Sibley, Wilbur, and the 1980s photo census (the Best Population Estimates), but both lineages of estimates indicated rapidly approaching extinction. The October survey counts paralleled the best estimates, but were consistently lower.

McLean-Sibley-Wilbur-1980s photo census lineage (our preference) — a continuing rapid decline of the species was apparent after 1950. By 1980 it was hard to avoid the conclusion that extinction might occur well before the end of the twentieth century, and it was clear that conservation efforts for the species had reached a final critical phase, with very little room for delay or error.

Condor Numbers around 1900

If we assume the validity of the 1940 to 1950 estimate of 150 birds, Sibley's estimate of 60 birds for the late 1960s, Wilbur's estimate of 30 birds for 1978, and the photo census counts in the 1980s, the wild population experienced a gradually increasing rate of decline over the years, with a 5 percent annual decline from 1950 to 1968, a 6.5 percent annual decline from 1968 to 1978, a 7 percent annual decline from 1978 to 1982, and a 17 percent annual decline from 1982 to 1985. Further, if the 5 percent annual decline estimated for 1950 to 1968 is extrapolated backward in time as a constant, it is conceivable that there could have been more than 1,000 Condors alive at the beginning of the twentieth century. However, the apparent trend toward increasing rates of decline in recent decades suggests this could be an overestimate, perhaps a strong overestimate, and we are not proposing this number as a reliable estimate of the population size in 1900.

In fact, whether the Condor population was substantially larger in 1900 than in 1950 has been a matter of some debate. Carl Koford (1953) proposed that the Condor population size of the 1940s had been stable for several decades. Although he did not supply persuasive supporting data for this assessment, it may have been based largely on the similarity of his estimate of 60 birds with the estimate of 60 birds given by Cyril Robinson for 1934 and the estimate of 40 to 100 birds given by Dawson in 1923. The estimate of Robinson was almost surely a major underestimate, however, based simply on the fact that it rested largely on high counts in only two limited geographic regions. The estimate of Dawson is not known to have been based on any systematic censusing efforts. Consequently, it would be unwise to conclude anything firm about population trends from the similarities in these estimates. Sibley likewise estimated 60 birds for the late 1960s, but he clearly had far fewer birds overall than Koford, judging from area by area comparisons of numbers of birds seen.

Ian McMillan (1968) went much further than Koford with a suggestion that the Condor population may have been very close to extinction in the early years of the twentieth century, after which it enjoyed a substantial increase. This conclusion was apparently based largely on an obvious local increase he noted in sightings in the vicinity of his ranch near Cholame in San Luis Obispo County in the mid-twentieth century.

McMillan's sightings were in what was by then a peripheral region of the species' foraging range, however, and could well have only represented a period of locally favorable food supplies that may have temporarily pulled in birds from other foraging areas, rather than being any indication of a general population increase.

Any population increase or stability through the early twentieth century must be seriously questioned because the numbers of birds being reported from a variety of counties, ranging from San Diego, to Orange, Riverside, Santa Clara, Santa Cruz, western San Benito, western Monterey, eastern Los Angeles, southern Ventura, and southern Santa Barbara, were apparently declining strongly in this period and because a little-known population of Condors resident in the Sierra San Pedro Martir of northern Baja California disappeared completely at this time. Although high counts of Condors in the Sespe and Sisquoc Sanctuary regions of Ventura and Santa Barbara Counties appeared to be holding strong, there were no clear signs of increase in these regions.

Koford (1953) did indeed mention several observers who believed that Condors were increasing on the foraging grounds in Kern County in the early twentieth century, and a number of relatively high flock counts of Condors were indeed reported there in the 1940s, but these counts, like increases in Condor sightings seen on the foraging grounds in northeastern San Luis Obispo County during the 1940s, might be more reflective of local shifts in food distribution or of increasing numbers of observers than indicative of any beneficial population changes.

It is well to remember that large congregations of Condors are an ephemeral phenomenon, especially on the foraging grounds, and the chances that large gatherings might be actually seen, counted, and reported by observers presumably depends in part on the numbers of observers frequenting Condor range. With a steadily increasing human population through the early twentieth century, apparent increases in the sizes of the highest counts in any region could occur in the absence of any actual increases in numbers of Condors. The important point to recognize is that in spite of increasing human populations, the numbers of Condors being reported generally declined in the great majority of regions in the early twentieth century, especially various breeding regions. Thus, it is hard to believe anything other than a progressive overall decline in numbers of Condors occurred through this period.

In the next two chapters we take a close look at causes of decline of the species and discover that mortality factors appear to have been the overwhelmingly dominant problem. Because of the naturally low reproductive potentials of the Condor, rapid increases in population size are highly unlikely to occur in the absence of major reductions in mortality factors. Yet there is no reason to suspect such reductions may have occurred in the early twentieth century, especially in view of the substantial number of Condors that were known removed from the population in that period by collectors, by continuing heavy shooting pressures on all wildlife, and by continuing predator poisoning campaigns. In fact, it is safe to assume that shooting and various kinds of poisoning, including lead poisoning, had been present since the first European settlers arrived in the range of the species. We strongly suspect that the combined effects of these and other mortality factors exceeded the reproductive capacities of the historic population on a continuous basis since early times. Although rates of population decline may indeed have varied from decade to decade, the steady, progressive shrinkage of the range of the species in the nineteenth and twentieth centuries gives good evidence that the overall population decline was probably also progressive and not interrupted by periods of population increase or population stability.

THE PERSISTENCE of any species through time depends on a favorable balance of reproduction with mortality. Long-term population declines can result either from abnormally low reproduction or from abnormally low survival rates, or from some combination of these problems. Low reproduction can trace either to a failure of enough adults to breed or to poor success in breeding efforts. Low survival can result from many different mortality factors and is an especially severe threat if it affects adults and if the species under consideration is one like the Condor with a naturally low reproductive rate, even under optimal conditions.

The primary challenge in rescuing a failing population of any species is to discover what is responsible for its decline. The task normally entails measuring rates of reproduction and survival and comparing these rates with the rates necessary to produce stable populations. Wherever significant deficiencies are found, their specific causes have to be traced and corrected. Only the correction of factors that are important in causing the decline can be expected to produce population recovery.

Unfortunately, little information was available on either reproductive or survival rates of the California Condor in historical times, and early hypotheses as to causes of decline were not usually framed quantitatively. Rather, discussions of causes of decline were largely limited to plausible generalizations about specific stresses that had been noted or suspected for the species, without any clear evidence that all major stresses had been identified and without clear evidence as to which of the postulated stresses might be of primary significance. Prior to the time the last birds were brought into captivity, conservation efforts proceeded largely on the basis of highly fragmentary knowledge of what the primary stresses to the species might be, and relatively little evaluation was made of the effectiveness of the conservation efforts that were implemented. In hindsight these efforts clearly proved unsuccessful in reversing the decline and may not have even appreciably slowed the decline.

A reasonably good understanding of the causes of the Condor's decline was not attained until the species had almost disappeared from the wild in the 1980s, and by that time it was much too late to attempt correction of the primary identified stress factors before the species would have been lost. The Condor was only preserved in the near term by removing it from a hostile wild environment and allowing it to rebuild its numbers in a more benign captive environment. Reestablishment of viable wild populations, however, still depends on correcting the most important stresses that caused critical endangerment in the first place, as there is no evidence to suggest that the main threats have fortuitously and spontaneously disappeared.

In this chapter we focus on the principal hypotheses that were advanced prior to 1980 regarding the causes of the California Condor's decline. These were the primary hypotheses to be critically tested in the greatly expanded research efforts of the 1980s, and only through a close look at these hypotheses can we provide a sense of the state of confusion and controversy that characterized historic Condor conservation efforts and an understanding of why subsequent research and conservation efforts proceeded in the directions that they did.

Missing from our discussion here is a consideration of lead poisoning, as this stress was not recognized in the species until after 1980 and did not figure in anyone's hypotheses or recommendations regarding conservation of the species before that time. The research methods used prior to 1980 could not have been expected to reveal this stress, even though it now appears plausible that lead poisoning may have been the most important stress in the long decline of the species. How this stress was ultimately identified and the evidence that now exists to implicate lead are part of the discussion in the next chapter, which mainly considers the investigations of causes of decline conducted by the intensive research program of the 1980s.

Shooting

Shooting was perhaps the most frequently hypothesized cause of decline for the Condor in early writings and was emphasized by Cooper (1890), Finley (1908a), Dawson (1923), Scott (1936b), Koford (1953), and Miller et al. (1965). Indeed, considerable historical documentation of shooting of Condors exists. Miller et al. (1965) and McMillan (1968) provided especially convincing accounts of shooting threats in the twentieth century, and in an extensive tabulation of historical reports of Condor losses, Wilbur (1978b) determined that museum and other purposeful collecting of Condors, mostly involving shooting, accounted for 177 of 301 known losses overall. Further, shooting accounted for 41 of the 64 losses where a cause was identified, and the cause was not museum or other purposeful collecting. On the surface, these figures suggested that a major fraction of historic mortality traced to shooting, although museum collecting had almost completely ceased by 1925.

Such a conclusion must be questioned, however, because of strong biases in the data. The 301 losses documented throughout history prior to 1978 surely represented only a tiny fraction of the total Condor deaths occurring during this period (likely less than 1 percent), and it is virtually certain that the recorded fraction of losses due to shooting was much higher than the actual fraction of overall losses due to shooting. By its very nature, museum collecting was a form of mortality that was very thoroughly documented in the historical record, regardless of what fraction of overall mortality it may have represented. Similarly, other shooting mortality was likely relatively well documented because (1) shot birds are often kept and displayed; (2) shooting is a cause of death that is relatively easy to diagnose because of obvious wounds, broken bones, and ammunition fragments in carcasses (pl. 51);

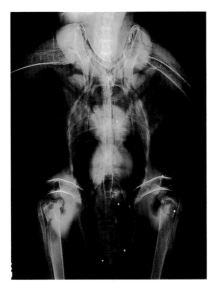

Plate 51. An X-ray taken of a living California Condor in 1985 indicated she had survived an earlier shooting incident but still had many shot pellets in her tissues.

and (3) birds abandoned by shooters are likely to be found by others because most shooting occurs near roads and trails.

Other sources of mortality were much less conspicuous and much less likely to be recorded than mortality due to shooting. For example, birds dying of disease or slow-acting poisons would unlikely either be found, correctly diagnosed, or reported in historical times. They could easily be completely missed in a historic tabulation, even though such mortality factors might conceivably be much more important overall than shooting. Therefore, although shooting was surely more than an occasional stress for the Condor population and was likely a substantial stress, unbiased data to demonstrate its true historic importance do not exist. Shooting was likely important enough to have contributed significantly to the continuing decline, but the extent to which it may ever have been the most important cause of decline is unknown.

Aside from collecting efforts, probably much of the historic shooting of Condors was motivated by curiosity or by the thrill that comes from firing at large flying creatures. In part, however, Condors were shot to obtain their large hollow wing feathers as containers for gold dust, a practice described by Taylor (1859b), Anthony (1893), Scott (1936a), Koford (1953), and Bidwell (1966) for a variety of regions and for dates as late as the 1930s. Although some authors have dismissed this practice as probably unimportant, the number of apparently independent accounts describing the practice, their wide geographic distribution, and the apparent prices paid for Condor feathers (one dollar per feather in one early account) suggest significant impact.

Clearly what was needed in the 1980s to attain an accurate assessment of the continuing role of shooting in the decline of the species was some relatively unbiased means of determining what fraction of continuing mortality might trace to shooting. Also needed was some means of determining whether the species was threatened primarily by mortality or reproductive factors.

Poisoning

Numerous early naturalists (e.g., Taylor 1859b, Henshaw 1876, Streator 1888, Cooper 1890, Bendire 1892, Finley 1908a, Lyon 1918, and Lopez in Latta 1976) also called attention to substantial mortality of Condors resulting from incidental poisoning associated with campaigns to rid California of large predators, especially Grizzly Bears *(Ursus arctos horribilis)* and Wolves *(Canis lupus)*. Many of these authors attributed the rapid decline of the Condor predominantly to this cause. The predator poisoning campaigns, mostly involving strychnine-laced carcasses, were sufficiently effective, when combined with other methods of control, that Grizzly Bears and Wolves were exterminated from the region by the 1920s. Nevertheless, predator poisoning campaigns continued through the rest of the twentieth century in an effort to control Coyotes *(Canis latrans).*

Unfortunately, very few of the early accounts of Condors being lost to such poisoning efforts were clearly firsthand (only Fry 1926), and because of this, a number of relatively recent authors, apparently following the lead of Dawson (1923), have discounted predator-poisoning campaigns as a major cause of the historic decline. This judgment has rested in part on an assumption that the Condor might be highly re-sistant to poisons because of its ability to tolerate rotten food and in part on an assumption that most published accounts of poisoning losses might be mere repetitions of early unsub-stantiated claims. In particular, Scott (1936b), Harris (1941), Koford (1953), and Wilbur (1978b) all minimized the poten-tial importance of early poisoning campaigns because of the paucity of firsthand documentation implicating this form of mortality.

Nevertheless, a general absence of firsthand accounts of poisoning incidents should not be surprising even if early losses of Condors to predator poisoning efforts may actually have been substantial. The people doing such poisoning were presumably quite uniformly ranchers in early days and pro-fessional, often government sponsored, predator killers in later years. Ranchers and professional predator killers had nothing to gain from publishing or otherwise publicizing the results of their efforts, and insofar as Condors might have been killed, they had something to lose, at least during the twentieth century. Especially under conditions where first-hand reports of poisoning Condors would carry likely penal-ties for potential authors, the validity of early accounts should not be judged simply on the basis of whether or not they were firsthand. Hearsay is not always erroneous, and when it is likely the only way one is going to learn about what is going on, it is prudent not to dismiss it too hastily. Further, a close look at the published early accounts of Condor poisoning strongly suggests that although they appeared to be mostly secondhand, they quite surely were not all simply repetitions of earlier accounts (see especially the accounts of Lyon 1918 and Lopez in Latta 1976).

Also, by 1980 there was more recent direct evidence of substantial risks to vultures from such predator-poisoning efforts. Miller et al. (1965) reported a case of three Condors evidently poisoned by strychnine in the carcass of a Coyote in 1950. One of the three died, and the other two were severely incapacitated but were eventually saved by provision of food and water over a period of time. Significantly this was also a case whose existence was deliberately suppressed by government personnel for many years. Another apparent strychnine-poisoned Condor was found at a Coyote bait in 1966 (Borneman 1966). These incidents, together with the virtual disappearance of Eurasian Griffon Vultures (*Gyps fulvus*) from Romania and Bulgaria, which was linked with strychnine poisoning of Wolves (Bijleveld 1974), raised strong doubts about the immunity of large vultures to strychnine and raised strong concerns as to the safety of Condors in regions of predator control using this poison. This position was later to be strongly supported by extensive data on strychnine poisonings of African vultures assembled by Mundy et al. (1992).

Strychnine was not the only predator poison of potential significance to the California Condor. Another was cyanide, and this toxin, rather than strychnine, was implicated as the primary threat to Condors in one of the reports of major losses of Condors in early times (Lopez in Latta 1976). As Lopez described matters for the Tejón Ranch,

> For a number of years before I arrived at the ranch [1874], Mr. Beale had employed Pete Miller to kill bears. This he did by shooting, trapping, and poisoning. In the mountains he hunted them, and also caught them in traps or deadfalls made of heavy timbers. When they killed sheep or cattle he put poison in the carcasses. Strychnine generally was used in

such poisoning but Miller used a special poison of his own. The nature of this poison he kept secret from everyone. It would kill everything: condors, buzzards, crows, ravens, foxes, coyotes, wolves, coons, wild cats, mountain lions, and bears. We even had dogs, valuable dogs, killed by this poison. Later I learned that Miller used cyanide of potassium. I was assured by everyone that before this poisoning was done, both wolves and condors were plentiful in the Tejón country.

Cyanide again came to be an important method of poisoning predators, in particular Coyotes, after the outlawing of compound 1080 (sodium fluoroacetate) in such efforts in 1972. As we shall see in the next chapter, one Condor was found dead of such cyanide poisoning in 1983 (pl. 52). Significantly, this was another case in which efforts were apparently made to conceal causes of the bird's death (see Snyder and Snyder 2000). How many other Condors may have perished from this cause in relatively recent times is speculative, especially in view of the possibility that additional cases could have been concealed or suppressed.

Plate 52. Poisoning incidents of wild California Condors have involved lead, strychnine, and cyanide, and no evidence exists for immunity of Condors to such toxins. This immature Condor was evidently killed by cyanide from a Coyote trap in 1983.

As a general view, there is no reasonable doubt that there were massive efforts to poison large predators in California, especially in the nineteenth and early twentieth centuries, usually employing strychnine- but sometimes cyanide-treated carcasses. Because there is direct evidence for toxicity of these poisons to Condors and no good evidence for immunity of Condors to these poisons (despite a widespread belief in this), it would be surprising if these campaigns were not killing substantial numbers of Condors as a side effect, regardless of the paucity of published firsthand accounts.

The certainty with which many early ornithologists identified predator poisoning as a major stress is notable, even if most presented only generalities about this threat. Conceivably, some might have offered more specific documentation if they could have predicted the degree of skepticism that would later surround their accounts. If early accounts claiming that hundreds of Condors were killed by such poisoning have any validity, this stress might well have been much more important than shooting overall. This was clearly the opinion of Charles Bendire (1892) who was on the scene and emphasized the near universal practice of poisoning predators on large California ranches. James Cooper, who was likewise on the scene, also mentioned poisoning before shooting in his 1890 account of the species' decline, and Lopez (in Latta 1976) evidently considered predator poisoning the major stress causing a decline of Condors in the Tejón Ranch region during the late nineteenth century.

Another poisoning threat discussed by Koford (1953) and emphasized by Miller et al. (1965) and McMillan (1968) was the potential loss of Condors to toxins used to kill California Ground Squirrels *(Spermophilus beecheyi)* in Condor range, especially compound 1080, a toxin introduced in the late 1940s. Koford and Miller et al. reported direct observations of Condors commonly consuming poisoned squirrels and kangaroo rats *(Dipodomys* spp.), so exposure of the Condor to such poisons is not in question. What was debated, however, was whether such exposure was severe enough to have killed or incapacitated any Condors. Neither Koford nor Miller et al. documented any adverse effects in Condors seen feeding on 1080-poisoned squirrels, and direct laboratory tests of the toxicity of 1080 suggested a high degree of resistance to this substance in the Turkey Vulture *(Cathartes aura)*, a close relative of the Condor, in contrast to high sensitivity in mammalian species (Ward and Spencer 1947, confirmed by Fry et al. 1986). Further, a direct telemetric evaluation by Hegdahl et al. (1979) of the threats of this substance in squirrel-control efforts in the field revealed no evidence for harm to raptors and vultures at the same time there was major mortality of squirrels and other small mammalian species. The overall data available in 1980 were not strongly suggestive that the 1080 used in squirrel-control campaigns constituted a major threat of immediate mortality to Condors. These data, however, did not rule out the possibility of significant detrimental sublethal effects of 1080 on Condors, especially if the substance were ingested over long periods.

Further, in addition to being used in squirrel-control efforts, 1080 was also used to some extent as a carcass-bait poison for Coyotes during the mid-twentieth century. Because the dosages of this material received by Condors from Coyote-control efforts were likely less well defined than in most squirrel campaigns, it is perhaps more probable that Condors may have been lost to 1080 used in Coyote control than to 1080 used in squirrel control.

Also worrisome was the use of thallium in squirrel-poisoning campaigns in the early to mid-twentieth century, as thallium is known to have killed some Turkey Vultures in the field (Linsdale 1931). Although no Condor deaths have ever been confirmed as due either to thallium or to 1080, the case remained open in 1980 as to some significant historical (and continuing in the case of 1080) impacts of these toxins.

Overall, Miller et al. (1965) argued that the decline of the

Condor was primarily linked to excessive mortality from shooting and poisoning (with special concerns about 1080 poisoning), and they suggested that reproduction of the species appeared to be proceeding normally. Their evidence for normal reproduction, however, was limited to documentation of what appeared to be relatively strong ratios of immatures to adults in some flocks and speculations about the total numbers of immatures and adults in the population. It was not based on any direct studies of nesting effort and nesting success. In the absence of accurate data on the total population size of the species and any data allowing systematic identification of individual Condors, age ratios seen in a few flocks can provide only a suggestive indication of healthy reproduction. Furthermore, the focus on 1080 poisoning as a potentially major factor in Condor mortality was based primarily on the finding of several dead Condors in a region where squirrel poisoning was practiced, not on finding 1080 residues in the dead birds. The causes of death of these birds could well have been other unidentified factors.

In sum, an accurate determination of the continuing role of various kinds of poisoning in the decline of the Condor was not possible prior to 1980. Comprehensive evaluation of the continuing role of poisoning rested on the development of means to obtain relatively unbiased data on the extent and causes of mortality. Although later evidence was to largely confirm the conclusion of Miller et al. (1965) that the major cause of decline was indeed excessive mortality, especially from poisoning, this confirmation was only achieved through the intensive studies of reproduction and survival of the species that became possible in the 1980s.

Food Scarcity

In contrast to Miller et al. (1965), Sanford Wilbur (1978b) argued that a principal problem of the Condor in the late 1960s and 1970s was poor reproductive effort, with very few birds attempting to breed. Wilbur documented apparently substantial recent declines in food supplies (for example, a decline of about a third in the number of Cattle *(Bos taurus)* in Ventura County between 1950 and 1970), and his failure to find many breeding pairs of Condors was, in his view, likely attributable to these declines. Wilbur emphasized, however, that the few birds that he knew were still breeding were showing good success in breeding attempts, and it is unclear how food stress might strongly affect breeding effort and simultaneously leave breeding success unaffected.

Unfortunately, the data to back up Wilbur's hypothesis were no more conclusive than the data summoned by Miller et al. to justify their opposing view. The primary data to suggest potential problems with breeding effort during the late 1960s and early 1970s were the low immature-to-adult ratios documented in the October Surveys of Condors during that period. Depressed age ratios were not apparent in other data for this period, however, as discussed in Snyder and Snyder (2000). In particular, the percent of Wilbur's Condor sightings that included immatures and the average number of immatures seen per sighting day by all observers showed no depression during the late 1960s and early 1970s. In addition, age ratios that can be calculated from Wilbur's estimates of total numbers of immatures and total population size yielded much higher immature-adult ratios than the October Survey.

Thus, it is unclear whether the October Survey data accurately reflected age ratios of the entire population. And even if low immature-to-adult ratios might have actually characterized the population at that time, they could just as easily have been caused primarily by fluctuations in age-specific mortality rates, nesting success, or both, as by depressed breeding effort.

Further, there was no persuasive direct evidence of pairs refraining from breeding during those years. Wilbur documented no clearly nonbreeding pairs and lacked the manpower to conduct a thorough assessment of use of potential nest sites. Nevertheless, he evidently believed that the relatively large flocks of Condors observed during his studies on the foraging grounds in southern Kern County (especially the Tejón Ranch) largely represented nonbreeders. Unfortunately, because Condors were not individually recognizable at that time, and the foraging ranges of individual breeders were not well known, it was alternatively possible that the concentrations of birds he observed on the foraging grounds might have been mainly made up of breeders. Although Wilbur documented very few fledglings during those years (approximately 1.5 per year), it was possible this was mainly a reflection of his limited coverage of potential nest sites.

Further, whether or not progressive declines in food supplies might cause a collapse in breeding attempts for the Condor depended crucially on whether or not the species was food-limited in the first place. The hypothesis that the species might be food-limited was testable by deliberately increasing food availability and seeing if this produced increased reproduction and population increases of Condors. Such efforts were begun in a preliminary way by Fred Sibley in the late 1960s and were continued much more systematically by Wilbur. Wilbur was particularly concerned that food supplies might be inadequate for the Condor population in summer and that an apparent decline in nesting activity in the Sespe Sanctuary region of Ventura County might be due to low food in this region. His attempts to test these ideas directly involved offering the birds supplemental carcasses in the vicinity of the Sespe Sanctuary on a throughout-the-year basis, coupled with efforts to document numbers of birds using the carcasses and efforts to find signs of nesting in the sanctuary.

The program of supplemental feeding of the population was carried on continuously through the 1970s, but, although the carcasses offered were indeed used with moderate frequency by Condors, there was no clear sign that the Condor population ever became strongly dependent on them. The numbers of birds using the carcasses never became very large, and use always remained highly sporadic. Further, there was no evidence for especially heavy use of the carcasses in the summer (use was characteristically strongest in winter and spring). Finally, and perhaps most importantly, the Condor population continued to decline strongly through this period (by about 50 percent overall), and no persuasive evidence was assembled for increased reproduction in the population. Taken together, these results failed to provide good evidence for food scarcity being a primary limitation or cause of decline for the species. Miller et al. (1965) likewise concluded that the species was not food-limited on the basis of direct estimates of numbers and seasonal distribution of carcasses available to Condors.

Clearly what was needed to further evaluate Wilbur's hypothesis in the 1980s was a major manpower-intensive effort to locate all nesting pairs and to determine their productivity directly, coupled with intensive study of all active pairs to determine if there were signs of food stress. In particular, studies of growth and development of Condor chicks would be one direct way to evaluate potential food limitations.

Human Disturbance of Nesting Areas

Although Carl Koford did not believe that the Condor population of the 1930s and 1940s was declining, he expressed major concerns in his 1953 Condor monograph about the potential impact of human disturbance on Condor nesting activity, suggesting that pairs could easily be caused to desert their nests by visits of humans to the near vicinity. These concerns formed a primary basis for Koford's support for the creation and maintenance of nesting sanctuaries free of human disturbance for the species.

Nevertheless, to our knowledge historic records fail to reveal a single well-documented case of Condors deserting a nest because of human disturbance per se (as differentiated from nest failures caused by the taking or killing of eggs, nestlings, or adults). Further, the amount of disturbance produced by repeated human visitations was quite substantial at a number of active nests, including the nest studied by Finley in 1906 and a number of nests repeatedly entered by Koford himself. Despite such disturbance, the adult birds at these nests were characteristically persistent in breeding activities.

Concerns regarding the potential effects of human disturbance reached an angry climax in a publication by Ian McMillan (1970), in which he claimed that the Condor population of the late 1960s had largely ceased reproduction as a result of nest entries by the U.S. Fish and Wildlife Service (i.e., by Fred Sibley). Unfortunately, McMillan had no comprehensive data on the amount of reproduction going on in the population, and he clearly underestimated it. Further, he failed to recognize that the great majority of Sibley's nest visits occurred at times when nests were not active and that of the five active nests Sibley visited, four were evidently successful in fledging young, with no good evidence that the entries caused either nest failure or failure of the pairs to attempt breeding in subsequent years.

Overall, the evidence in 1980 for the idea that disturbance from nonpredatory human visitations to nests constituted a major threat to Condor reproduction was weak, although there had unquestionably been a certain amount of predation on nests by humans for purposes of egg-collecting and young-collecting for zoo exhibition. Only a couple dozen young were ever taken for exhibition, however, and in view of a known ability of many species of birds to lay replacement eggs after egg removals, the impact of the egg collectors on California Condors may have been much less than might at first appear. Only about 80 Condor eggs were ever collected in historical times, and it seemed very possible that many of these eggs might have been followed by replacement eggs laid in the same season by the affected pairs.

To achieve a better understanding of the potential effects of human disturbance on nesting Condors, studies of the 1980s would need to carefully document by distant observations of active nests both what constituted normal reproductive behavior in the species and what levels of response were observed to a great variety of naturally occurring disturbance factors. Special attention needed to be paid to the issues of whether or not the species might be capable of replacement egg laying and whether disturbance around nests might ever cause nest desertion.

DDE Contamination

In the decades immediately following World War II, a variety of carnivorous birds came under major stress from contamination with newly developed chlorinated hydrocarbon pesticides used to control noxious insects. Attention became focused in particular on the effects of DDT, or more specifically DDE, the chemical to which DDT transforms in living systems. DDE was shown to cause eggshell thinning and resulting high rates of egg breakage and nesting failure in species such as Peregrine Falcons *(Falco peregrinus)* and Brown Pelicans *(Pelecanus occidentalis),* and it was associated with strong population declines in these species. It was reasonable to suspect it might also be having negative impacts on the California Condor.

This concern became crystallized in a paper of 1979 in which Lloyd Kiff and his colleagues documented massive (32 percent) eggshell thinning highly correlated with substantial DDE contamination in a small sample of Condor eggshell fragments collected in the 1960s and 1970s. Combined with Sanford Wilbur's (1978b) low estimates of Condor reproduction for the late 1960s and 1970s, these data suggested that the

Condor might be a classic, and potentially extreme, example of the DDE syndrome. In fact, Robert Risebrough (1986) was to claim subsequently that "it appears that the California Condor would have been the first species to become extinct as a result of DDT use, had the use of this chemical not ended in time."

Nevertheless, the sample size of eggshells analyzed in the Kiff et al. (1979) study was very small, and it was not yet clear if the Condor population had been suffering from increased egg breakage and poor nesting success—the most usual detrimental effects of DDE on bird populations. Although the 32 percent shell thinning strongly suggested the potential for serious reproductive effects, it was not proof in itself of such effects. Further research was needed to more thoroughly examine the correlation of DDE with shell thinning and to carefully examine reproduction of the species, both past and present, to determine if there were clear signs of reproductive stress correlated with levels of contamination. Because DDT was not introduced until after World War II, it obviously could not account for Condor declines prior to that time. Conversely, even though uses of DDT were largely terminated in 1972, this substance might still have been having negative impacts on Condors in the 1980s, as it is a highly stable and persistent material in biological communities.

Collisions

Evidence for collisions with overhead wires and objects as a significant source of mortality goes back to Koford (1953), who documented a Condor death to an apparent collision with a surveyor's stake (pl. 53) and a collision of a Condor with a fence wire that left it injured and led to its shooting by a rancher. In addition, a Condor was directly observed in an apparent fatal collision event with overhead wires in the Sierra Nevadas in 1965. Although no early researchers suggested

Plate 53. Carl Koford with a dead Condor that was the apparent victim of a collision with a surveyor's stake in 1941.

collisions as a primary cause of decline, by 1980 collisions were being found increasingly to be a significant mortality factor in studies with other large birds (see summaries in Olendorff and Lehman 1986, Brown et al. 1987, Brown 1993). Presumably, collisions had been an increasing threat to Condors over the years with the proliferation of utility lines, and as we shall see in chapter 9, it is noteworthy that many collision deaths of Condors were documented in early releases of Condors into the wild in the 1980s and 1990s (pl. 54). The collision events documented for these birds, however, may have been unnaturally frequent because of abnormal attractions of the released birds to human dwellings and other structures. Historic Condors showed no known tendencies to perch on power or telephone poles or to land on buildings or other human structures.

As with shootings and poisonings, the principal research means for documenting the importance of collisions in the life equation of the Condor in the 1980s was anticipated to be

Plate 54. Collisions, sometimes associated with electrocutions, have been a significant source of mortality in recent California Condor releases and caused the death of this bird recovered near Fillmore in 1993.

detailed studies of mortality of the species, especially through radiotelemetry. The random and fortuitous documentation of collisions, shootings, and poisonings in the historical record proved only that these were sources of mortality, not necessarily important sources.

Calcium Stress

As discussed in chapter 2, Condors, like other large vultures, have a strong need for supplementary calcium in their diet, and severe calcium stress had been documented in certain populations of Cape Vultures *(Gyps coprotheres)* in South Africa that were deprived of normal bone supplies by the extirpation of Spotted Hyenas *(Crocuta crocuta)* (Mundy and Ledger 1976). Even earlier, Cowles (1958) similarly postulated calcium problems for the California Condor as a result of shifts in diet to large mammal carcasses that might result from fire suppression in chaparral-dominated areas. In his view, absence of fire over long periods would lead to declines in rabbits *(Lepus californicus, Sylvilagus audubonii)* and other small mammals from which Condors might obtain more bone materials than from Horses *(Equus caballus)*, Cattle, and Mule Deer.

A scarcity of calcium could lead to problems such as bone breakage of nestlings, as seen in the Cape Vulture of South Africa. Nevertheless, Cowles presented no direct evidence that Condors were suffering from calcium deficiencies, and as already discussed in chapter 2, there has been only one known case, in 1939, of bone breakage by a nestling Condor, and this could have resulted from an accident during banding or from some other accident, rather than from calcium deficiency. Enough Condor nestlings had been observed over the years to make it unlikely that chronic bone-breakage problems might have been missed by observers, and although Cowles's hypothesis had an attractive logic to it, it suffered from a lack of supporting data. Elucidating the importance of calcium stress in the 1980s would rest on expanded observational studies of active Condor nests.

Habitat Loss

No species can exist without adequate amounts of the habitat to which it is adapted, and it is a fundamental truism of modern conservation that the great majority of endangered species are threatened mainly by habitat loss or modification. That the California Condor might be suffering from habitat loss or disturbance had been a popular concept for many years, especially as advanced by Koford (1953) and Miller et al. (1965), and historic conservation efforts for the Condor over the years were heavily slanted toward creation of Condor preserves, as we shall see in chapter 7. Clearly the Condor had witnessed major changes in land use practices in recent times, especially in its foraging areas. Nevertheless, whether these changes had been an important cause of decline was still uncertain in 1980. These changes had little impact on the historic nesting areas of the species mainly because of the remoteness and ruggedness of the nesting regions. Similarly, no one had documented any obvious shortages of roost sites or bathing and drinking sites in the nesting regions. Instead, if the species had been suffering from habitat losses, it appeared almost certain that the important losses must have been on the foraging grounds, where significant areas of pasturage had been converted to urban development and agricultural uses over time and where total numbers of domestic animals had declined substantially over the years.

Perhaps the primary way in which habitat changes might have negatively affected Condor populations was through diminished food supplies. It was not clear, however, that the Condors were suffering from food scarcity. If the species were being limited by food supplies, some evidence of this should have been apparent in problems with nesting success and nesting effort. Yet Wilbur's efforts to subsidize the wild population with food through the 1970s yielded no obvious lessening of the rate of population decline, and surely no population increase. In fact, as discussed in the preceding chapter, the best available population estimates suggested a progressive increase in rate of population decline during this period. These features of the decline did not suggest habitat loss or modification as the primary driving force. In historical times, the Condor clearly ranged over a huge diversity of habitats, and there is no good evidence of it being a habitat specialist, beyond a potential need for relatively open areas in foraging and a need for adequate nest sites in regions relatively free of natural enemies such as Golden Eagles (Aquila chrysaetos).

The primary means envisioned for further evaluating the potential importance of habitat limitations in the 1980s were studies of feeding ecology and nesting ecology of the species.

Any signs of poor growth of nestlings or depressed nesting effort or success might ultimately trace to such habitat limitations, and an absence of problems in these areas would suggest that habitat constraints were not the most important threats to the species.

Other Miscellaneous Stresses

In addition to the stresses discussed above, historical data suggested a number of other stresses with potentially significant negative effects on the species. The large number of prehistoric Condor remains found in the La Brea tar deposits of Los Angeles, and occasional recent records of drownings and mirings tabulated by Koford (1953), suggested a chronic threat of such difficulties, especially in regions of surface oil deposits. Presumably, Condors attracted to oil pools ran a risk both of mirings and of oil toxicity, although it was very difficult to assess the magnitude of threat overall on the basis of historical information. Evaluating the magnitude of a continuing threat of this nature in the 1980s depended on detailed research into causes of continuing mortality.

Another historic stress that may have been of demographic importance was the regular use of Condors in sacrificial ceremonies by Native American tribes (see account of Friar Boscana in Harris 1941). Just how many Condors may have been lost to such practices is highly speculative, but historical accounts suggest that the ceremonies were widespread and potentially of a magnitude adequate to produce population effects, as indicated by Harris (1941). Apparently these ceremonies continued to some limited extent until the mid-nineteenth century (see account of Heerman 1859), but they were undoubtedly declining strongly by that time along with massive declines in Native American populations. We agree with Harris that the sacrificial ceremonies of early times may have provided a significant stress for local Condor popula-

Plate 55. The Wheeler fire of 1985 incinerated 184 square miles of Condor nesting habitat. Although this fire did not harm any active Condor nests, similar fires in the past have undoubtedly caused the failure of some nests, and one nest of 2003 would surely have been lost to fire if it had not already failed from other causes.

tions, although Simons (1983) argued against this point of view. No evidence existed to suggest any continuing losses of this sort by 1980.

Still another stress of potential importance was the fires that occasionally sweep through the chaparral habitats where most Condors have nested (pl. 55). Some preflight Condor nestlings and eggs have undoubtedly been lost over the years to the intense heat and smoke of such events. However, free-flying birds have presumably been able to avoid such threats, and it was doubtful in 1980 that the frequency of fires had ever been high enough to constitute a major threat to nests. To our knowledge, there is no historic record of a Condor nest ever failing to fire, and the first known case where a nest would surely have failed to this cause only appeared in the re-cent release program in 2003 — at a site that had already failed earlier from other causes (see chapter 9). In the 1980s and subsequently, intensive nesting studies offered a direct means for evaluating the magnitude of this threat.

Thus, historical data and hypotheses as to causes of the Condor's decline varied greatly, but by 1980, persuasive evidence was still lacking for the primary importance of any specific stress factor in the decline of the species. By this time it was even still unclear whether the Condor's difficulties lay primarily in reproductive or mortality stresses, and there was an overriding need for better evidence on causes of decline.

Perhaps the three best-known and most-plausible historical hypotheses about major causes of decline were those of Miller et al. (1965), Wilbur (1978b), and Kiff et al. (1979). Miller et al. saw the Condor's decline as primarily a result of excessive mortality caused by shooting and various sorts of poisoning, and they believed there was no good evidence of reproductive deficiency. Wilbur, in contrast, felt that recent problems lay mostly in a failure of most adults to attempt breeding, potentially caused by food scarcity. Kiff et al. attributed the recent decline in large part to DDE contamination, a stress that generally operates through eggshell thinning, increased egg breakage, and reduced nesting success. Unfortunately, none of these authors had assembled the sort of evidence needed to rigorously demonstrate the validity of their hypotheses, and without such demonstrations, attempts to correct the causes of decline still risked missing the most important threats. No comprehensive documentation of either reproductive or mortality rates was achieved until the 1980s, when all Condors became identifiable as individuals through photo census efforts, when intensive radiotelemetry studies were initiated, and when massive efforts to find and study all nesting pairs were made.

BY THE LATE 1970s, conservationists became acutely aware of the rapidly approaching extinction of the California Condor. Whether the species should simply be allowed to disappear or whether a much-enhanced effort to save it should be mounted finally reached the level of a national debate. With the apparent rate of decline there was no time for further delay, and largely thanks to the persistent efforts of the National Audubon Society, a consensus decision was reached among a variety of federal, state, and private agencies that the species should not be abandoned without a major last-ditch effort.

In 1980, the U.S. Congress provided funding for greatly expanded Condor conservation activities, which allowed a major increase in the number of people and organizations studying the species. A cooperative program was initiated in that year, involving the U.S. Fish and Wildlife Service (USFWS) and the National Audubon Society as principal research partners but also with important contributions from the California Department of Fish and Game, the U.S. Forest Service, the Bureau of Land Management, the Los Angeles Zoo, the San Diego Zoological Society, and several local museums and universities. This expanded program resulted in large part from the recommendations of a special panel convened in 1978 by the American Ornithologists' Union and the National Audubon Society to review the Condor program. This panel urged initiation of captive breeding of the species to ensure its survival, along with greatly intensified research efforts on the wild population. With increased personnel, it finally became possible to evaluate many of the factors potentially causing the decline of the species.

Intensive efforts to improve understanding of the Condor's decline involved attack on a number of fronts: (1) efforts to improve censusing of the population and to identify individual Condors so that mortality rates could be accurately assessed, (2) efforts to locate all nesting pairs and to make quantitative analyses of breeding effort and success, and (3) radiotelemetric efforts to study Condor movements and to locate dead Condors to determine specific causes of mortality. By the mid-1980s, sufficient progress was achieved in all these components that it was finally possible to identify the important causes of decline with reasonable confidence. The most important stress ultimately identified—lead poisoning—was, ironically, not a stress that had been identified earlier by Condor researchers.

Censusing Efforts and Determination of Mortality Rates

In 1981, the research program began a collaborative effort with Eric Johnson and his students at California Polytechnic State University in San Luis Obispo to explore the possibility that each Condor in the population might be recognizable as an individual on the basis of its pattern of flight-feather molt. This research soon established that Condors go through feather replacements at a fairly slow rate during a molting season that runs from early spring to late fall. Further, the patterns of replacement of the eight fingerlike primary feathers at the end of each wing (primaries numbered three through 10) were especially easy to see and document (pl. 56). Each year a Condor replaced about half of its primaries, and the specific feathers replaced varied greatly from bird to bird and also from one wing to the other wing in a single bird. Each feather had a lifetime of about two years before being replaced, so a bird's pattern of molt in one year was generally close to the opposite of the previous year's pattern. Thus, a bird molting primaries three, six, seven, and 10 on a wing in one year could be predicted to molt primaries four, five, eight, and nine on the same wing in the next year, whereas its molt pattern on the other wing was often completely different but likewise alternating from year to year. Like fingerprints, these feather molt patterns turned out to be unique with each individual, and when coupled with unique patterns of feather

Plate 56. Flight photographs of California Condors reveal distinctive patterns of molt and feather damage that allow reliable identification of individuals through time. This individual was first identified in 1982 and died of lead poisoning in 1985. At the time of this photograph, primaries six and 10 on the bird's right side and seven and 10 on the bird's left side were short and still growing. Note also the broken-off tips to primary seven on the bird's right side and primary eight on the bird's left side.

damage (various nicks, holes, and broken tips) that were visible on most all birds, they allowed the reliable identification of all individuals in the population on a continuing basis through the years.

To make such an identification system work, all field personnel carried cameras with powerful telephoto lenses and tried hard to photograph all flying Condors they encountered, carefully noting the date and location of each photograph. Each primary feather that was lost during molt took about four months to regrow, and during this period, there was first a gap visible in the array of feathers, then a steadily lengthening new feather in the gap until it reached full size. This growth rate was sufficiently slow that it was not difficult to take enough photos of individuals to keep track of their molt sequences. In practice, birds photographed even just once a month provided far more information than was needed for accurate identification and determination of molting patterns.

Crucial to the success of this effort was the discovery of a centrally located observation point in the foraging range of the species where nearly all birds in the population could be photographed with some frequency (pl. 57). This observation point was known as The Sign and was a roadside pulloff at a large wooden U.S. Forest Service sign that overlooked some of the best ranchland foraging habitat in southwestern Kern County. Approximately 25 percent of all flight photographs taken came from this one location, and these formed the core photographs for the analyses that were made (Johnson et al. 1983). Additional photographs were taken in all nesting areas and at other strategic locations throughout the foraging range.

Fortunately, the great size of the Condor and its characteristic soaring flight allowed informative photographs of wing patterns from long distances. Whenever birds were within several hundred yards of observers, and observers took care to

Plate 57. Crucial to the success of photo-census work of the 1980s was close monitoring of a roadside observation point called The Sign, which overlooked important foraging habitat in southwestern Kern County. All birds in the remnant population used this foraging region to a greater or lesser extent.

Plate 58. The senior author (left) and Eric Johnson (right) sort Condor flight photographs in the first thorough photo census of 1982.

accurately focus their cameras, it was possible to get photographs of sufficient quality to allow clear identifications.

For analyses, four-inch by five-inch black-and-white enlargements were made of each photograph, and each was labeled with the appropriate date and location (pl. 58). These enlargements allowed side-by-side comparisons with other photographs to speed the identification and differentiation of individuals. A file was kept of the photographs of each bird, arranged in time sequence through the years. As new photographs were obtained, they were compared with the files on the known birds until an appropriate match was found for the date in question. Matches during the winter nonmolting period were based largely on unique damage patterns in flight feathers. To guard against any errors in identifications, the entire flight-photo files were ultimately checked independently by three different researchers, all of whom confirmed the accuracy of the bird identifications.

Thus from 1982 on it became possible to identify all birds in the population, no matter where they were photographed. This allowed a determination of the range and movements of all individual birds. When photographs of an individual bird ceased abruptly, it was generally because the bird in question had died or was moribund, and the photographs allowed determination of the approximate time of the bird's disappearance. With this kind of documentation, it finally became possible to accurately census the population on a continuing basis and to calculate mortality rates with considerable precision.

The photographic censusing efforts confirmed a precipitously declining wild population, starting with an apparent 23 birds in early 1982, dropping to 21 birds in late 1982, to 19 birds in late 1983, 15 birds in late 1984, and just nine birds in mid-1985. Three birds were trapped into captivity in mid-1985, and after another bird died in early 1986, most of the remaining birds were taken captive in the same year. The last wild bird was captured in the spring of 1987.

Mortality rates proved to be very high. Between early 1982 and early 1986, when the population dropped from a probable 23 birds to five birds, the average annual mortality rate was 26.6 percent—or 18.9, 16.7, 43.2, and 27.5 percent for the four years, respectively. As Meretsky et al. (2000) later calculated, expanding on earlier demographic analyses by Mertz (1971) and Verner (1978), these mortality rates were far in excess of what a stable population of normally reproducing Condors might tolerate. Specifically, under conditions of normal reproduction, the population could be expected to withstand no more than about 9.9 percent annual mortality and still remain stable. Thus the Condor population of the mid-1980s exhibited a mortality rate more than two and a half times as great as could be sustained by a normally reproducing population. Clearly a major cause of decline in this period was excessive mortality. Meanwhile, no clear evidence for significant reproductive problems was emerging, as is examined in the next section.

Surprisingly, the average annual mortality rate for immature birds (22.2 percent) was slightly lower than the rate for adults (26.8 percent), suggesting that the main factors responsible for mortality in the species were not markedly age-dependent. In most bird species, mortality rates for immatures greatly exceed the rates for adults, because immatures are more susceptible than adults to most mortality factors. As we shall see in the discussion below, the near equality in mortality rates of immature and adult Condors is one of the more suggestive pieces of evidence in evaluating which mortality factors were most important in the species' decline.

Reproductive Studies of the 1980s

Breeding Effort

Because of a crucial need to know whether Condors were still attempting to breed with normal frequency and success, many of the increased numbers of personnel of the 1980s were allocated initially to checking of historically known nest sites for activity, and then increasingly to following Condors in the field to find new nest sites. Only two of the 25 nest sites observed active with eggs or young in the 1980s turned out to be previously known sites, and all the rest were new sites—that is, sites never before documented by anyone (although evidence such as layers of internal excrement clearly showed that most had also been active in earlier years). Thus, for the purpose of documenting reproductive activity, the efforts to follow Condors to their nests proved far more productive than the efforts to check historically known sites.

No nest-finding efforts on the scale mounted in the 1980s had ever been made by earlier Condor researchers. Although

Carl Koford and Fred Sibley had found moderate numbers of nests in the late 1930s to mid-1940s and in the mid- to the late 1960s, respectively, both these researchers worked mostly alone, neither focused on following birds to nests, and neither was able to achieve more than partial coverage of the range within which Condors were known to nest. Thus, these researchers lacked the resources, personnel, and methodology needed for a thorough assessment of Condor reproduction, and most of the nests they found were located by following up on earlier reports of Condor activity in various locations. The Condor program remained minimally staffed through the 1970s, and only small-scale efforts were made to find and monitor nests during that decade. In contrast, more than 12 individuals were devoted full time to nest finding and monitoring by the mid-1980s.

By 1983 and 1984 it became possible to combine nest-finding efforts with nest-monitoring efforts for essentially all pairs in the population. Efforts to observe active nests were generally conducted on a continuous daylight basis, something that had not been accomplished previously, although Carl Koford (1953 and unpublished notes) had studied one pair on an approximate 30 percent coverage basis from March 23, 1939, to March 24, 1940. The knowledge gained through continuous observations at nests of the 1980s allowed direct and accurate determinations of rates of nesting success and causes of nesting failure. Further, they allowed an evaluation of the efficacy of earlier nest-finding efforts and led to significant refinements in methods for finding active pairs.

In particular, from the relatively infrequent exchanges of adults documented at active nests, as described in chapter 3, it became clear that to safely eliminate any candidate nesting area as being active during the season pairs might have eggs (roughly January through June), researchers had to watch the area for about five to six consecutive days, as no external activity (flying Condors) was seen at some active nests for periods as long as this and occasionally longer. Things were more favorable in the nestling period (roughly July onward) when Condors exhibited much more activity around nests. But even at this latter season, researchers often did not see any activity in the vicinity of an active nest through a full day of observations and sometimes not for two straight days. Thus from July to the end of the year it was generally advisable to watch each candidate nesting area for three consecutive days before concluding the area was almost surely inactive.

Prior to the 1980s, the extremely infrequent visits of Condors to their nests in many pairs was not known to researchers (Koford [1953] documented unusually frequent visits of adults to the one nest he studied closely), and to our knowledge no historic researcher had ever observed a prospective nesting area with the intensity and duration needed to prove it was inactive. Thus an observation process adequate to rule out nesting activity in any area was developed only in the 1980s. Although moderate numbers of nests were found active in earlier studies, these instances mainly involved cases where observers were fortunate enough to coincide in their visits to nesting areas with visits of the adults, where observers accidentally flushed birds from their nest caves, where observers fortuitously came across late-stage nestlings or fledglings outside their nest caves, and where observers actually entered historic nest sites and found eggs or chicks or their remains. To our knowledge, during the early years of Condor study the only individual to regularly follow the procedure of patiently tracking adult Condors to their nests (the method on which we came to rely) was Kelly Truesdale, an egg collector who focused his attentions on one pair of Condors in San Luis Obispo County in the early 1900s.

Unfortunately, the general absence of thorough nest checking before 1980 means that one can conclude very little about the total amount of Condor nesting activity going on in historical times, even in well-known nesting areas. Simply visiting an area for a few hours to watch a historic or prospective nest from a distance—the usual methodology followed in the 1970s, for example—could not reliably reveal whether there was nesting going on in the vicinity, as pairs often do not exchange places at nests for many days and birds attending nests between exchanges are often invisible within their nest sites from any distance. Further, checking limited to known historic sites would leave many active sites undiscovered simply because these nests were in locations different from the known historic sites. Even when active sites were located reasonably close to historically known sites, they were often inconspicuous sites unmarked by obvious physical signs, such as external excrement, that might permit their detection as active sites without actually seeing the birds entering or leaving the sites.

The only way to be sure about nesting activity in candidate nesting areas was to watch them for flying Condors for many consecutive days from strategic lookouts that allowed for good viewing of large areas of habitat, often in remote locations that were difficult to reach (pls. 59, 60). Even once the presence of a pair was positively confirmed in a nesting region, it was often necessary to change observation points repeatedly to finally be in a position to see the birds go to their actual nest site. In view of the rugged terrain in which the Condors were nesting, this effort often involved many additional days of work, progressively homing in on the pair's center of activities.

Just how difficult it sometimes was to find active Condor nests, even when their general locations were strongly suspected, is well illustrated by the case of the CV pair of 1983. The approximate location of the nesting territory of this pair of Condors (neither of which was radiotelemetered until 1984) was first identified during May 1982, when the pair was discovered associating with an apparent recent fledgling at a roost near the top of an escarpment about midway between the Sespe and Sisquoc Sanctuaries. Starting in mid-February 1983, intensive efforts involving many observers were made to find the active nest site of this pair by locating promising

Plate 59. Hunts for nesting pairs of Condors in the 1980s were primarily conducted from strategic lookouts that covered large areas of suitable habitat. Here Jack Ingram scans for Condors from a lookout in the Sespe Sanctuary in 1982.

Plate 60. Because of winter snows, certain high-elevation nesting regions could only be reached with difficulty in the early breeding season. Here Brad Bush sets off on a 1985 expedition to check one of the giant sequoia *(Sequoiadendron giganteum)* groves in the Sierra Nevada for nesting activity.

overlooks of canyons in the vicinity of where the birds had been seen roosting and by maintaining long watches from these overlooks. Unfortunately, although the pair was occasionally seen from various overlooks through spring 1983, it took until May 18 of that year before the actual nest canyon was finally found by a distant sighting of what appeared to be a classic nest exchange centered on the canyon. In this observation, a single Condor flying into the area was joined in soaring above the canyon by a second Condor rising out of the canyon. The second bird then departed the area while the first bird dropped down into the canyon.

Significantly, this very canyon had been closely watched earlier (February 14–15 and March 27–29) without seeing any evidence that it was the nesting canyon, and it had been tentatively eliminated from consideration. Efforts through April until early May were focused on many other canyons in the region, and it was only when these efforts also proved unproductive that another try was made with what proved to be the actual nest canyon. The nest site, now containing a nestling Condor, was a small and extremely well-hidden cave in a remote narrow gorge that could only be viewed directly from a nearby observation point that was far from any established road, trail, or lookout point. Not surprisingly, there is no record of the site ever having been visited by Condor researchers earlier or identified historically as a nest site, although layers of excrement inside the site indicated clearly that it had been used multiple times in the past.

Thus, the investments in personnel needed to find all Condor nests to make a comprehensive determination of the amount of nesting activity in the population were substantial, and these investments continued to be necessary even when observations could be aided by such techniques as radiotelemetry of breeding birds. Any direct measures of breeding effort in the population were simply far beyond the human resources available prior to 1980. Nor were there any reliable indirect methods for making such assessments prior to the

1980s, as there were no means available at that time for identifying individual Condors to make comprehensive counts of numbers of pairs and juveniles in the population. Nevertheless, it finally proved to be logistically possible to make reliable direct assessments of breeding effort in the early to mid-1980s, largely because the Condor population by then had dropped to relatively few pairs plus a few assorted nonbreeders that were all identifiable through photo-census data, and enough observers were ultimately available to cover all active pairs on a reasonably thorough basis.

In Table 1, summary data on nesting effort are presented for the Condor population for 1982 through 1986. Data were too incomplete for 1980 and 1981 to include these years and too few birds were left in the population in 1985 and 1986 (including only a single female in 1986) for these years to be very useful in calculations. The data for 1985 were also compromised by extremely heavy mortality of breeding pairs just prior to the start of breeding. The data from 1982 through 1984, however, probably give a reasonable picture of representative levels of breeding activity in the population.

From 1982 to 1984 the percentage of adults that were members of egg-laying pairs varied between about 50 and 80 percent. On the basis of percent of known pairs that laid eggs, the average for these years was very close to 80 percent, a percentage that was almost identical to comparable averages for apparently healthy populations of four species of large African vultures studied by Mundy (1982). Judging from this similarity, there was no obvious sign of any abnormal depression of breeding effort in the California Condor during the early to mid-1980s.

Available data also suggest that breeding effort may have been healthy in the late 1970s. In particular, observations of early 1981 (Snyder 1983, Johnson et al. 1983) revealed the existence of at least eight different dark-headed Condors (one- to three-year-olds) in the wild population. In tracing back how many pairs it would have taken to produce these

TABLE 1. Breeding status of wild California Condors in the early breeding season

	1982	1983	1984	1985	1986
Total Condors	23	19	15	10	5
Breeding pairs	4	4	5	2	1
Nonbreeding pairs	2	1	0	0	0
Unknown status pairs	1	1	0	0	0
Unpaired male adults	0	0	2	2	3
Unpaired female adults	1	1	1	2	0
Unsexed unpaired adults	2	1	0	0	0
Immatures	6	5	2	2	0
Percent adults paired	82	86	77	50	40
Percent adults breed	47–59	57–71	77	50	40
Percent pairs breed	57–71	67–83	100	100	100

Source: Adapted from Snyder and Snyder 1989.

Note: From 1983 onward, all eggs and nestlings were removed for the captive breeding program. Mortality of five adults over the winter of 1984–85 disrupted four breeding pairs; the male of a new pair in 1985 was lost early in the breeding season, and although no egg was laid the pair was ranked as a breeding pair because it was known to be heterosexual and copulations were proficient. Two pairs in 1982 (one of them also present in 1983) did not perform proficient copulations and were strongly suspected to be homosexual pairs of males.

juveniles, given what is known about Condor mortality rates, reproductive rates, and reproductive success, it is possible to calculate that the eight juveniles implied the likely existence of at least six, and more probably seven or eight, egg-laying pairs in the three preceding years. This many egg-laying pairs accounted for the great majority of adults presumably alive during the period, however, as there also were at least two subadults and the eight juveniles in 1981 and the total population of Condors was only about 30 individuals in 1978 and perhaps only about 25 to 27 individuals in 1980. Thus, it was unlikely that there was any substantial number of nonbreeding adults at the time. In late 1982, there were still seven dark-heads and subadults in the population, which then consisted of 21 known individuals, a fraction also compatible with both normal breeding effort and success in the population.

For years earlier than the late 1970s, data were simply insufficient to venture any calculations of percent of adults breeding, but as discussed in chapter 5, there are no persuasive grounds to suspect chronic deficiencies in reproductive effort in these earlier years. The apparently strong ratios of immatures to adults in many of the flocks seen by Miller et al. (1965) in the early 1960s and Koford (1953) in the 1940s do not suggest problems in this sphere, although the data are too incomplete to allow firm conclusions about immature-to-adult ratios for the full populations in existence during those periods. Immature-to-adult ratios were also strong on the October Surveys of the mid-1960s and mid-1970s, and the only period during which they appeared deficient on the surveys was in the late 1960s and early 1970s. As discussed in chapter 5, however, other evidence suggested much higher immature-to-adult ratios for this period, and low ratios, even if true, could be produced by factors other than low breeding effort. In any event, the late 1960s and early 1970s represented at most a very brief period of low ratios that cannot be taken as good evidence for any chronic problems in breeding effort for the population.

Breeding Success

As examined in detail by Snyder and Snyder (2000), the data on reproductive success, like those on reproductive effort, do not suggest any major problems in breeding during the 1980s. For 17 nesting attempts found in the early stages and allowed to proceed at least until the late nestling stage or to failure or imminent failure, either seven or eight can be considered likely successful, for a nest success rate of 41 to 47 percent. This figure represents a nest success rate similar to that documented for several comparable species of solitary-nesting vultures in Africa by Mundy (1982) and that documented for Black and Turkey Vultures (*Coragyps atratus* and *Cathartes aura,* respectively) in North America by Jackson (1983).

Rates of nesting success for Condors of the 1960s and of the 1930s and 1940s were apparently similar to those of the 1980s, although sample sizes of analyzable nests were relatively small for these other periods. If we limit analyses to nests whose fates would likely have been determined from the egg stage, whether they failed or succeeded, there were between four and six successful nests out of 12 (33 to 50 percent) in the 1960s and between two and five nests successful out of nine (22 to 56 percent) in Koford's (1953) studies of the 1930s and 1940s. Fred Sibley (1968) similarly calculated a 45 percent success rate for Condor nests of the mid-1960s, and Carl Koford assumed a roughly 50 percent nest success rate in his own nesting calculations.

Thus nesting success apparently averaged between about 40 and 50 percent through the entire period of nesting studies from the late 1930s through the 1980s. Like the rate for the 1980s alone, this rate of nesting success appears to be within the typical range for solitary-nesting vulture species, although it is less than has been generally documented for colonial nesting griffon vulture species.

The primary cause of nesting failure documented in the 1980s was apparent predation on eggs by Common Ravens *(Corvus corax),* with Ravens involved as either a contributing or ultimate cause of the failure of five nests and a potential cause of two other failures. Circumstantial evidence suggested that Ravens could also have been involved in some of the failures of the 1940s and 1960s, as detailed in Snyder and Snyder (2000), but direct observational evidence was lacking for causes of almost all of these earlier failures. Ravens may have been an increasing problem for Condor nesting success in the 1980s as a result of apparently increasing Raven populations in southern California, and this species has apparently been the main cause of egg breakage in the species. Most nest failures of the Condor in all historical eras have involved egg breakage, with very few losses of nestlings documented.

Intensive studies of nests during the 1980s revealed no signs that breeding pairs were having difficulty in procuring enough food for successful reproduction, as already discussed in chapter 3. However, planned efforts to study growth and development of Condor chicks were aborted after the accidental death of a chick during handling in mid-1980. This incident was one of the important factors in delaying the initiation of radiotelemetry, as is discussed in the mortality section below.

Summary of Reproductive Studies

Overall, the intensive studies in the 1980s of reproduction of the remnant wild population indicated reasonably good breeding performance. Although the population was experiencing some moderate difficulties with Raven predation on eggs, most adult Condors were breeding and nest success was within the usual limits seen in other populations of solitary-nesting vultures. Concentrations of adults on the foraging grounds in the San Joaquin Valley foothills during the breeding season were found to be made up primarily of breeders, not the nonbreeders postulated by Wilbur (1978b). Further, the rate of decline seen in the wild population was far too rapid to be explained primarily by reproductive deficiencies, and even if nesting success may have been hypothetically lower in the mid- to late twentieth century than in earlier centuries, as a result of increasing Raven populations, it was still reasonably strong and not an obvious major problem for the species. Finally, there was no evidence that habitat limitations might be adversely affecting reproduction or that reproductive effort was being negatively affected by human disturbance. In fact, pairs proved to be exceedingly regular and repetitive in breeding efforts despite extensive egg-removal operations, involving considerable disturbance, carried out in the mid-1980s (see chapter 8).

The reproductive results were especially important in evaluating the potential importance of DDE contamination, as DDE normally produces population declines in birds by reproductive effects and DDE had been suggested as a primary cause of the Condor's recent decline, as discussed in the previous chapter. Considerable attention was given to gathering relevant data to assess this hypothesis in the 1980s, primarily by monitoring DDE levels and shell thinning in eggs (pl. 61) and by documentation of reproductive performance of the population. The data accumulated, as reviewed by Snyder and Meretsky (2003), provide no persuasive evidence that DDE had important impacts on the species, either in the 1980s or at the height of the DDE era in the 1960s. In fact, the decline of the species evidently even accelerated in the 1980s without any clear signs that DDE contamination was a major cause. No strong correlation of DDE with shell thickness was confirmed in a large sample of shells, and the decline, at least from the late 1970s through the 1980s, was clearly traceable

Plate 61. To assist studies of potential effects of DDE contamination, Dave Ledig collects Condor eggshell materials from a nest active in 1982.

mainly to mortality effects, not reproductive effects. The apparently severe eggshell thinning documented for the 1960s may have been primarily an artifact of the small size of the eggs being sampled at that time, although size was not measured for any egg in that period. Suggesting this possibility is the fact that there was no apparent concurrent increase in egg-breakage rates and no sign in any era that thin-shelled eggs were any more susceptible to breakage than thicker-shelled eggs. A very strong correlation of shell thickness with egg size exists for the Condor (as it does for all birds in general), and shell thickness per se reveals little about eggshell strength without also knowing what size eggs are under consideration.

Mortality Studies of the 1980s and Later Years

With mounting evidence that the wild Condor population was still breeding normally, and with steadily accumulating data on survival rates of individuals achieved through photo-census methods, it became apparent by the mid-1980s that the primary difficulty faced by the species was excessively low survival (that is, high mortality), not low reproduction. Actual mortality rates in the 1980s averaged more than 25 percent per year — about two-and-a-half times as great as might be tolerated by a stable population with the reproductive rates exhibited by the Condor in the 1980s. Significantly, mortality rates were as high in adults as in immatures, an unusual situation for bird species in general and a situation with especially grave implications. With mortality rates this high, no conceivable reproductive rate could have allowed population stability or increase, even if one assumed the highest possible reproductive performance of the species (100 percent nest success and all adults breeding at maximal frequency).

Thus, determining exactly what was causing the excessive mortality of the population became the primary research need of the program, and here, as discussed in chapter 5, the historical data on shooting, poisoning, and other stresses were much too fragmentary and biased, as well as potentially too out of date, to give any reliable accounting of principal mortality threats still affecting the population. What was needed was a means of determining the most important continuing causes of mortality by finding dying or dead birds in the wild with much greater frequency and reliability than could be achieved by fortuitous discovery of dead birds. Only by comprehensive necropsies of a relatively substantial and unbiased sample of such birds could the comparative importances of various mortality factors finally be expected to become clear. To this end, the primary hope for success lay in radio-tracking as many birds in the wild population as were possible to capture and fit with transmitters (pls. 62–64).

The radiotelemetry program for Condors was under direction of John Ogden of the National Audubon Society and

Plate 62. The first Condor trapped for radiotelemetry in 1982 is provided with wing-mounted transmitters by the senior author, together with Bruce Barbour, Jesse Grantham, and Pete Bloom. This bird perished from lead poisoning a year and a half later.

Plate 63. John Ogden releases a radioed Condor from a sky kennel in 1984.

Plate 64. Nine California Condors were fitted with patagial radio transmitters during the 1980s. This individual, first captured in 1984 and still subadult at the time of this photograph in 1985, was later to become the last surviving California Condor in the historic wild population before his recapture in early 1987. This Condor again became a breeder in the wild in 2004.

was based mainly on trapping free-flying birds with a cannon net that had been developed specifically for vulture research by the Vulture Study Group in Africa. In cannon-netting of Condors, a large camouflaged net was fired explosively over birds that had been attracted to feed at a carcass bait. Some birds were alternatively trapped using a pit trap, in which a trapper in a hidden subterranean pit reaches up to grab the legs of a bird feeding at a carcass. Both trapping methods often involved long waits, sometimes for many days or weeks, before Condors came to baits.

Once they were captured and immobilized, birds were fitted with radio transmitters mounted on the leading edges of their wings, an attachment design that had been field tested with Andean Condors *(Vultur gryphus)* in Peru by Mike Wallace (Wallace and Temple 1987b). The radios used varied from solar-powered to various battery-powered units with long lifetimes and ranges up to a 100 miles and more, and each Condor normally carried two radios, one on each wing, to reduce the chances of losing contact with the bird through transmitter failure. Radios used in the 1980s for California Condor were developed by Bill Cochran and Gene Bourassa.

Unfortunately, radiotelemetry of Condors was not initiated until late 1982, in part because of vigorous opposition to this technique from those who feared that the risks of such operations might exceed the benefits and in part because of the loss of a Condor chick in a 1980 handling accident unrelated to radiotelemetry. Ultimately only nine birds in the historic population were ever radioed, and many of these birds were radio-tagged for only relatively short periods of time before a decision was made (in late 1985) that all birds should be taken into captivity. Largely because of these factors, few mortalities were ever documented by the telemetry program.

The low numbers of dead birds recovered in the 1980s, however, was also due to the fact that a disproportionate fraction of the birds dying proved to be birds that were not wearing radio transmitters. This bias was especially pronounced in

Plate 65. The first documented Condor to have died of lead poisoning was recovered through radiotelemetry in the Sierra foothills in March 1984. John Schmitt examines the corpse under an apparent roost perch in a tall pine from which the bird had apparently fallen.

the winter of 1984–85 when five of seven unradioed birds were lost, whereas only one of eight radioed birds perished. Although this result at least suggested that the radios themselves were not a major cause of mortality, it was an unfortunate result from the standpoint of data that were not obtained on causes of mortality.

In sum, 11 of 15 Condors dying between 1982 and 1986 were never recovered and their causes of death remain unknown. Nevertheless, four birds dying in this period were recovered (two by radiotelemetry and two by fortuitous discovery) and the causes of death were determined for all four. Further, despite the small sample size, the specific causes of mortality determined for these birds gave a sufficiently compelling picture of the nature of mortality threats faced by the species that the conservation program was obliged to completely reorient its priorities and strategies. Three of the four

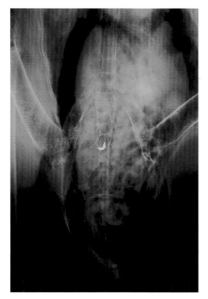

Plate 66. An X-ray of the lead-poisoned bird of 1984 revealed a small bullet fragment in its digestive system. A liver sample from this bird contained 35 parts per million lead wet weight, a concentration high in the toxic range.

Condors recovered were victims of apparently independent lead-poisoning events, presumably all resulting from ingestion of lead ammunition fragments in carcasses (two of the three birds still had ammunition fragments in their digestive tracts). These were the first wild Condors ever diagnosed with lead poisoning, but there was no reason to believe that this was a new problem, considering the apparent source of the contamination (pls. 65, 66). And because the three cases appeared to be independent events, it quickly became plausible to hypothesize that lead poisoning might be and might have been the most important factor stressing the life equation of the Condor.

The three cases of lead poisoning in the mid-1980s, together with the evidence for massively excessive mortality rates in the population, were sufficient to convince first the California Fish and Game Commission and later the USFWS

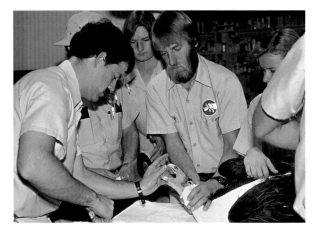

Plate 67. Veterinary personnel of the San Diego Zoological Society attempt to save another Condor suffering from acute lead poisoning in late 1985. The poisoning of this bird was the major factor that prompted the U.S. Fish and Wildlife Service to authorize trapping of all the remaining wild Condors.

that the remnant Condor population faced an apparently overwhelming mortality threat from lead that could not be quickly countered (pl. 67). Both organizations ultimately agreed that the only prudent course of action was to capture all the remaining wild Condors to at least preserve the species in captivity. This eventually came to pass despite strenuous opposition, including an unsuccessful lawsuit mounted by the National Audubon Society. Audubon's opposition was based primarily on a fear that removing all birds from the wild would doom habitat protection efforts for the species.

Evidence implicating lead contamination as a major threat to the species has only increased in subsequent years and has come from a number of sources. Perhaps the most convincing support has come from the numerous cases of acute lead poisoning occurring in releases of captive-bred Condors into the wild in the 1990s and 2000s, despite provision of a food subsidy of lead-free carcasses in all release efforts (see chapter 9).

Because released birds have routinely carried radio transmitters, it has been possible to get much more information on mortality threats in releases than was obtained for the historic wild population of the 1980s.

Provision of clean carcasses was the primary justification presented for doing releases without actually removing the threat of lead poisoning in release regions, but it has proved only partly successful in preventing such contamination. By 2000, at least five released birds had died from lead poisoning and another 16 were only barely saved from acute lead poisoning by emergency chelation therapy—a therapy that removes lead from the bloodstream and allows it to be excreted. In addition, other released birds had disappeared without being recovered and necropsied, so there could well have been additional lead-poisoning mortalities that went undocumented.

Since 2000 still more cases of acute lead toxicity have continued to occur, both in California and Arizona (by the end of 2002 a total of more than 30 birds had been given emergency chelation therapy), and it now appears clear that lead contamination is truly a widespread, common, and chronic threat. Presumably the incidence of lead poisoning in the release program would have been much greater if the birds had not been provided with regular clean food, and thus the recent cases of severe contamination represent only a very minimal documentation of the full threat still present in the wild. Despite the presumed reduction in mortality achieved by the clean-carcass subsidy, lead has been far and away the primary threat of mortality documented in the release program when acutely poisoned birds saved by emergency chelation therapy are counted as effective mortalities, as they should be (see Meretsky et al. 2000, 2001; Beissinger 2002). The importance of the lead threat to Condors has been further confirmed in a review by Fry and Maurer (2003).

Another piece of evidence consistent with the importance of lead poisoning is the fact mentioned earlier that mortality rates of immatures were no higher than mortality rates of adults in the historic wild Condor population. As free-flying Condors of all ages presumably eat about the same amount of food, and as food is the presumed main source of lead contamination in the species, one might predict that all ages would be equally susceptible to this threat. For most other mortality threats, for example shooting and collisions, one would predict much higher susceptibility of juveniles than of adults.

The probable equal susceptibility of all-aged Condors to lead threats is strongly supported by the lack of age differences found in lead contamination of another scavenger in Condor range—the Golden Eagle *(Aquila chrysaetos)*—as studied by Pattee et al. (1990). The Pattee et al. study was also valuable in documenting another feature of lead contamination that is consistent with it being a major stress to Condors—the seasonal pattern of contamination. Contamination of Golden Eagles was greatest in the fall through winter period—exactly the season when most Condor mortalities of the 1980s had occurred—and the season that was the main hunting season in the region. Overall, Pattee et al. found that 36 percent of the 162 Eagles they trapped in Condor range had elevated lead levels, and that although the extent of contamination varied from region to region, lead-contaminated birds were found in all regions sampled. The Pattee et al. study was not designed to document the extent of mortalities of Eagles to lead contamination (only survivors were sampled), but such mortalities have been noted in other studies.

The incidence of lead contamination was also high in another carrion-feeding species of the region, the Turkey Vulture. In a study by Wiemeyer et al. in 1986, 10 of 16 Turkey Vultures tested from Condor range had elevated lead levels in their bones. Further, five of 14 wild Condors sampled for blood-lead concentrations between 1982 and 1986 had elevated lead levels. Two of the five (IC1 and SBF) subsequently died due to lead poisoning and a third disappeared without being recovered.

Together, these data indicate a major mortality threat to Condors from lead contamination of carrion food supplies, presumably tracing to the ammunition fragments that remain in many carcasses of animals killed by shooting. In many, but not all, cases of lead-poisoned Condors, ammunition fragments were still present in the digestive tracts of the birds when they were recovered dead or dying, and it is clear that both shot pellets and bullet fragments were contributing to the problem. Although lead shot was replaced with steel shot for waterfowl hunting on federal lands in the late twentieth century, this did not affect the use of lead shot for hunting of terrestrial species, and it is the latter that Condors have been much more likely to consume. No nontoxic replacements for lead bullets are yet in wide use for hunting of the species most commonly killed with bullets, such as Mule Deer *(Odocoileus hemionus)*, and evidence so far strongly suggests that Mule Deer carcasses not recovered by hunters have been among the most important sources of lead contamination faced by the species.

Lead poisoning was not the only mortality threat detected in the 1980s and during early releases. One of the four dead birds recovered in the 1980s was a victim of cyanide poisoning at a Coyote *(Canis latrans)* trap (pl. 68). This is the only modern case known of such poisoning, although it is possible that others had occurred but had not been reported. As discussed in Snyder and Snyder (2000), it is very difficult to evaluate the historical extent of threat from this source, but since California banned the use of cyanide in Coyote traps in 1998, this poison has presumably no longer been a threat in California since that time. No cases of cyanide poisoning have been detected in Condor releases between 1988 and the present.

Next to lead poisoning, the second-most severe mortality threat in releases through 1999 was collisions with overhead wires, accounting for seven deaths. Whether the incidence of this threat might have been similar in the historic wild population is questionable, however, as the released birds have ap-

Plate 68. Marilyn Anderson (left) and Art Risser (right) prepare a Condor for X-ray analyses at the San Diego Zoo in 1983. This bird turned out to have died from cyanide poisoning.

parently been especially susceptible to collisions because of their abnormal attractions to humans and human structures (see discussion in chapters 9 and 10). Although it is reasonable to rank the collision threat as low to moderate for the historic population, it is not clear that it was more than this. Its importance in releases has been declining strongly, almost surely because of training of pre- and post-release birds to avoid perching on dummy utility poles fitted with electroshock mechanisms.

Of great interest is the fact that only a couple of shooting incidents of Condors were documented in the release program through 1999, a relatively low rate that suggests this threat may well have been substantially overrated in importance in the past. Although shooting is clearly a continuing threat (additional birds have been lost to shooting since 1999), data do not suggest it is presently a major threat, although it may have been a greater threat in the past.

Other miscellaneous causes of mortality in released birds

have included predation by Golden Eagles and Coyotes, drowning in a slippery-sided pothole, and poisoning with anti-freeze, but none of these threats appears to represent a major mortality factor comparable to lead poisoning in magnitude. For a full tabulation of all causes of mortality in releases through 1999, consult Meretsky et al. (2000).

Importance of the Lead Poisoning Threat

Thus, recent data on mortality threats to Condors in the wild strongly suggest lead poisoning as the principal threat. Although it is clearly not the only threat, lead poisoning has evidently been enough of a threat that it is extremely doubtful that self-sustaining wild Condor populations can be achieved in the absence of success in removing this threat. Unlike some of the other threats to the species, lead poisoning is a threat that can be removed with some efficiency, but this has not happened as yet. Why not is something we consider mainly in chapter 10.

Lead poisoning is hardly a threat confined to Condors, and as mentioned above, it was enough of a threat to waterfowl to have led to replacement of lead ammunitions with steel ammunitions in hunting of such species on federal lands by the early 1990s. Nevertheless, lead poisoning continues to kill many waterfowl to this day, largely because of spent lead ammunitions in the substrates of many wetlands. For example, lead poisoning appears to be the major cause of mortality today for some Trumpeter Swan *(Cygnus buccinator)* populations. Other species at substantial risk include other scavengers and predators of hunted animals, such as Bald Eagles *(Haliaeetus leucocephalus)* and Golden Eagles (Clark and Scheuhammer 2003), although Condors may be particularly at risk, in part because they cast pellets (including ammunition fragments) much less frequently than do most raptors. Condors may also have a propensity to ingest ammunition fragments because their chronic needs for supplemental bone in the diet may lead to preferential intake of small hard objects encountered in carcasses.

In addition to posing a threat of mortality, lead contamination also poses other potential risks to wildlife. In humans, significant mental deficits, both in learning and IQ levels, have been traced to lead concentrations far below lethal levels, and it is reasonable to speculate that such deficits may also be characteristic of Condors that are subjected to similar levels of contamination. The discovery of such problems in humans has been a major force leading to the progressive elimination of lead in gasoline, plumbing, and paint. Lead ammunition represents another source of such contamination in humans, primarily through the consumption of lead particles in the flesh of hunter-shot game. Thus, the value of removing lead ammunitions from the market extends to many species additional to the Condor, including humans. With the recent development of nontoxic replacements for lead ammunitions that have hunting properties equal to that of lead, there is every reason to pursue, and no persuasive reason to avoid, the complete elimination of this threat.

With respect to human health, it is not clear that there is any truly safe level of lead exposure other than zero exposure, and there is no evidence that minute quantities of lead might have any beneficial biochemical functions. The same may well be true for Condors and other species. Although no careful research has yet been carried out with potential sublethal effects in Condors, it seems probable that birds surviving sublethal exposure to lead may nevertheless be significantly compromised in their capacities to deal with other challenges and may subsequently suffer enhanced mortality rates and other negative effects that are difficult to trace conclusively to lead contamination. For example, frequency of collisions of Condors with overhead wires could plausibly rise as a function of deficits in coordination due to the nerve damage produced by low levels of lead. Elevated lead has indeed been documented in swans colliding with overhead wires in the United Kingdom.

As an overall view, the studies of the 1980s and data from more recent releases of Condors into the wild are strongly supportive of the general position of Miller et al. (1965) that the primary causes of the Condor's decline were mortality factors. However, the main mortality factor found in the 1980s and subsequently—lead poisoning—was not a specific factor considered by Miller et al. (1965) or any other historical Condor researcher, and it is highly questionable that even full correction of the mortality factors that were discussed by Miller et al. and others would have been enough to have reversed the decline. This is not so much a criticism of Miller et al. and other workers as it is a clear indication of the need for comprehensive research on the factors limiting endangered populations. Neither Miller et al., nor Wilbur, nor Sibley, nor Koford had anything like the sorts of resources necessary for such a task, and the crucial need for such research was not even generally recognized until the late 1970s. Even then, the value of intensive research was vigorously disputed by extreme elements of the conservation community.

Comprehensive understanding of limiting factors is a general need for all endangered species recovery programs, not just the Condor program, and attempting to conserve endangered populations without such understanding is often an exercise in futility. One simply cannot hope to conserve populations suffering mainly from lead poisoning by saving habitat, stopping direct shooting of Condors, or preventing various other kinds of poisoning, and yet the latter were in fact the directions in which historic efforts to conserve the Condor were aimed—with no signs of a resulting recovery. This is not to suggest that these were negative activities and not worth doing, only that they were insufficient to the task and in some respects a diversion from much more crucial needs.

AS ONE OF THE most awesome symbols of pristine wilderness in North America, the California Condor has inspired conservation efforts that date back nearly a century. Although clearly viable wild populations of the species do not presently exist, it is encouraging that after a very close brush with extinction in the 1980s, the Condor still endures as an apparently vigorous captive population. Truly viable wild populations of the species may not have existed at any point in the past two centuries, but there is no compelling reason to believe that they cannot be achieved in the future. If they are ever achieved, it will be a credit to the efforts of literally many hundreds of people over many decades and will represent the culmination of one of the most exhausting, drawn-out, and expensive species-conservation efforts ever attempted.

In this chapter, we review the history of the Condor conservation efforts that preceded the intensive efforts involving captive breeding and reintroduction that began in the 1980s and are underway today. Many sincere and dedicated conservationists fought bitterly against today's approaches, yet these approaches ultimately became the only means left with any hope of success when earlier worthy efforts failed to produce a recovering population. Although some observers attributed the continued decline of the species through the mid- and late twentieth century to a failure to properly implement historic conservation approaches, it is hard to see how the historic approaches would ever have solved the lead-contamination problem that appears likely to have been the primary driving force of the historic decline.

Habitat Protection

Outside of general protective laws prohibiting shooting and collecting of the species dating back to the turn of the twentieth century, the first major steps on behalf of the Condor were measures to protect significant regions of nesting and roosting habitat in the mountains of southern California. Two sanctuaries were created on National Forest lands that were known to host substantial numbers of Condors, primarily as nesting-roosting populations. The first of these was the Sisquoc Condor Sanctuary in Santa Barbara County (pl. 69) that was established in 1937, largely through the efforts of Robert Easton, a Santa Barbara businessman, and Cyril Robinson, the deputy supervisor of the Los Padres National Forest. Although relatively small, only 1.9 square miles, this sanctuary was the location for some of the largest concentrations of Condors seen in the 1930s, and it was still frequented to some extent by Condors almost to the end of the wild population's existence in the mid-1980s.

A much larger 54-square-mile Condor sanctuary was established in Ventura County in 1947—the Sespe Condor Sanctuary (pl. 70)—largely as a result of Carl Koford's intensive research in the region. This sanctuary was believed by many to hold the key to the salvation of the species because many of the Condor nests that had been found were located

Plate 69. One of the major assembly points for Condors in the 1930s was Sisquoc Falls, a central feature of the Sisquoc Condor Sanctuary. Condors gathered here for drinking, bathing, and nesting.

Plate 70. The Sespe Condor Sanctuary of Ventura County was the main site of Condor research and conservation efforts from 1939 through the 1970s. The sandstone cliffs of this region provide numerous caves for Condor nesting but are also associated with petroleum deposits. The process of fully protecting this area was a struggle against competing economic interests.

within its boundaries. Unfortunately, the very sandstone cliffs that were so suitable for Condor nesting in the Sespe were also a repository for natural deposits of petroleum, so from the very start there were competing pressures for exploitation of the region's mineral resources. Full protection of the sanctuary was not achieved until 1971, when it was fully withdrawn from mineral exploitation and by which time it had grown to 82.8 square miles in size. Many additional habitat protection measures have been enacted since creation of the Sisquoc and Sespe Sanctuaries, and most of these are listed in the timeline provided at the back of the book.

Major protected areas of Condor range under the jurisdiction of the U.S. Forest Service and the National Park Service are illustrated in map 4. These areas include various wilderness areas, Sequoia National Park, and other sanctuaries, and together they make up a very large portion of the known his-

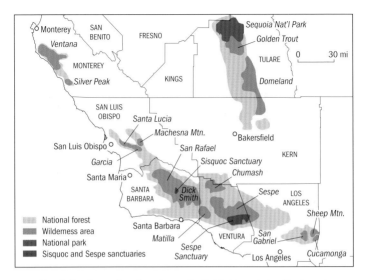

Map 4. A major portion of condor nesting habitat is on federal lands and is protected by various sorts of reserves of the U.S. Forest Service and the National Park Service.

toric nesting region of the species. In addition, some significant areas of the historic foraging range have now been acquired for conservation, most notably the Hopper Mountain National Wildlife Refuge (formerly the Percy Ranch), immediately adjacent to the southern border of the Sespe Sanctuary, and three reserves in southwestern Kern County—the Bitter Creek National Wildlife Refuge (formerly the Hudson Ranch), the Wind Wolves Preserve (formerly the San Emigdio Ranch), and the Carrizo Plains (pl. 71). Although the extent of the foraging range now dedicated to conservation purposes includes only a fraction of the recent foraging range, it is a substantial and still-growing acreage. Perhaps the most discouraging aspect of continuing habitat-protection measures of historic range has been the launching of development plans for the Tejón Ranch, a huge private holding comprising

Plate 71. The San Emigdio Ranch, an important foraging area for the historic Condor population in the southern San Joaquin Valley foothills, became the Wind Wolves Preserve in 1996.

the Tehachapi foraging zone of south-central Kern County, traditionally a pivotal feeding region for the species.

Other Historic Conservation Measures

Beyond habitat conservation, a variety of other conservation measures were proposed by Koford (1953) and Miller et al. (1965), including (1) federal protection for the Condor, (2) increased education and law-enforcement efforts to reduce shooting and molestation of Condors, (3) increased protection from human disturbance for nesting and roosting areas, (4) further study of the effects of poisons (especially 1080) on surrogate species, and (5) a full-time Condor warden to patrol areas of high Condor activity. In addition, these workers called for (6) an end to Coyote *(Canis latrans)* control using poisoned carcasses and (7) closure of some specific Condor flyways to shooting, especially the Sierra Madre Ridge in Santa Barbara County and Pine Mountain in Ventura County.

It is important to emphasize, however, that neither Koford, nor Miller et al. (nor any other more recent agency involved with Condor conservation) recommended a cessation of all shooting in the range of the Condor. Then, as now, such a step would have been politically explosive and very likely counterproductive to Condor conservation. Likewise neither Koford nor Miller et al. (nor any agency more recently involved with Condor conservation) pressed for an immediate ban on use of 1080, probably for similar reasons. Compound 1080 was indeed banned for use in Coyote control in 1972 and for use in ground-squirrel control in 1985, but not because of any good evidence that it was affecting Condors.

Most of the recommendations of Koford and Miller et al. were ultimately implemented. The Condor was given federal protection under the Endangered Species Acts of 1966 and 1973; greatly increased education efforts on the plight of the species did occur through publication of numerous magazine articles and through innumerable talks and discussions, especially those given by John Borneman of the National Audubon Society; protection of the Sespe Sanctuary was enhanced by seasonal firearms bans in buffer areas and by closure of outlying nesting areas such as the Piru Gorge; and private inholdings in the Los Padres National Forest, such as the Hi Mountain nesting area were purchased and given protection. Furthermore, a full-time Condor warden (John Borneman) was appointed by the NAS in 1965; the Sierra Madre Ridge Road was kept closed to public vehicles; aircraft were banned from the Sespe Sanctuary region; and an intensive study of the field effects of 1080 poisoning of California Ground Squirrels *(Spermophilus beecheyi)* was conducted by the U.S. Fish and Wildlife Service (USFWS) in 1977 (but indicated no clear threats to raptors or vultures—Hegdahl et al. 1979, 1986). In addition, Coyote trapping by the Animal Damage Control (ADC) branch of the USFWS became limited to scent-baited traps.

Miscellaneous Conservation Efforts of the 1970s and Early 1980s

As already described in chapter 5, and as is described in chapter 9, there were two major attempts to aid Condors by providing them with a supplemental food supply. The first of these attempts was conducted in the 1970s in an effort to increase reproduction in the wild population; the second was conducted in 1985 in an effort to reduce lead-contamination threats in the wild. Neither feeding program resulted in strong dependency of wild Condors on food subsidy, and neither was clearly successful in achieving stated goals.

A variety of other steps were taken to enhance the prospects of the species by the expanded program of the 1980s. These included a variety of repairs made to defective nest sites and especially efforts to prevent predation by Common Ravens

Plate 72. Efforts to protect Condor nests under severe threat from Common Ravens sometimes involved removal of the Ravens. The success achieved here by Bruce Barbour, however, was only temporary, as a new Raven quickly replaced one removed and the Condor egg under threat was ultimately lost to a Raven.

(*Corvus corax*) on Condor eggs. The latter involved attempts to remove or discourage Ravens that were seen directly threatening Condor nests (pl. 72). Results were mixed, with good success in protecting one nest from Ravens and a lack of success in others. Ravens proved to be difficult to approach and remove, and even when removal was successful, replacement Ravens often appeared rapidly so that the threat remained. Plans were under development in 1984 to attempt to condition local Ravens to avoid Condor eggs by taste-aversion methods (see Nicolaus et al. 1983), but these plans were put aside when almost all Condor pairs were lost during the winter of 1984–85 and the program was plunged into a crisis ultimately resulting in capture of all the remaining wild birds.

Summary of Early Conservation Efforts

Despite the many protective steps taken prior to the early 1980s, the decline of the species continued and in fact even appeared to accelerate. In hindsight, the failure of early actions to reverse the decline is not surprising. For even if all the recommendations of Koford, Miller et al., and others had been perfectly implemented, the chances are still very great that the decline of the species would have continued, as none of these recommendations offered an effective remedy to the major problem identified in the 1980s—lead poisoning. Nor did they consider the probably significant impacts of deaths due to collisions; nor did they offer more than a partial solution to the shooting threat.

Fortunately, as was discussed in the preceding chapter, the response to the failure of the actions taken was ultimately a widespread consensus that much better data were needed on the causes of the Condor's decline and a general recognition that the species would likely perish without the aid of captive breeding.

CHAPTER 8
CAPTIVE BREEDING

WITH THE CAPTURE of the last wild California Condor in spring 1987, conservation of the species became totally dependent on the success of captive breeding. Yet by that time, the species had never been bred in captivity, and it was not clear that such efforts would be successful. It was known that the species' closest relative, the Andean Condor *(Vultur gryphus),* was quite amenable to captive propagation, but it is never safe to predict that any species will thrive in captivity just because of success with its close relatives. In an ideal world it would have been much wiser to have experimented with captive breeding of California Condors long before the program finally became fully committed to this technique by a lack of any viable alternatives. Many species just do not adapt well to captivity, and the approach could have turned out to be a dead end.

Fortunately, the California Condor has indeed proved to be a successful breeder in captivity, and the captive population, under intensive management by the Los Angeles Zoo, the San Diego Zoological Society, the Peregrine Fund, and, most recently, the Oregon Zoo, has grown enormously in just a few years. Although the goal of viable wild populations has not yet been achieved, the threat of immediate extinction of the species has been greatly reduced.

Early Opposition to Captive Breeding

Captive breeding came very late in conservation efforts for the Condor, primarily because this approach was long and effectively opposed by important parties in the conservation community. Carl Koford (1953) was the most conspicuous early antagonist, perhaps mostly because of his fear that captive breeding could displace efforts to conserve the wild population, but he mainly expressed grave doubts that captive-reared Condors could be successfully reestablished in the wild and fears that releases of captives into the wild could introduce zoo diseases into wild populations. In his opposition, Koford had firm allies in Alden Miller, the McMillan brothers, and the National Audubon Society. Together, these parties and their followers were successful in preventing the establishment of a captive program for more than three decades after it was first formally proposed.

The first proposal to begin captive breeding of the species came from Belle Benchley (pl. 73) of the San Diego Zoological Society (San Diego Zoo). Benchley and K. C. Lint of the society had had great success during the 1940s in breeding Andean Condors in captivity, and as a natural extension of this success, Benchley approached the California Department of Fish and Game in 1949 with a proposal to take a pair of California Condors from the wild. The goal was not just to breed them in captivity but to serve also as a source of birds for release into the wild in an effort to bolster the wild population.

Plate 73. Belle Benchley, former director of the San Diego Zoological Society, was the first person to organize efforts to initiate captive breeding of California Condors.

The Department of Fish and Game did not oppose captive-breeding efforts, in part because it believed that taking of two birds from the wild posed no significant threat to a wild population it believed numbered about 150 birds. With the support of the department, the California Fish and Game Commission (CFGC) approved Benchley's proposal in 1952, although it expressly forbade releases of Condor progeny into the wild, cautious about the risks of inadvertently introducing diseases from captivity into the wild.

Approval of the permit, however, led to a major campaign of protest by opponents. This campaign successfully bypassed the commission's decision by persuading the California State Legislature to forbid any taking of the species from the wild after 1954. By that time, the Zoological Society of San Diego had not yet been successful in capturing any wild California Condors, so not only was the immediate prospect of captive breeding averted, opponents had also forestalled the authorization of any similar proposal that might be advanced by any organization in the future. It was only many years later, in 1971, that the state prohibition of any removal of Condors from the wild was changed to allow capture, as for captive breeding, under permit.

Important to the early success of opponents to captive breeding was a doomsday scenario presented by Alden Miller in 1953 that the taking of two Condors from the wild could represent a major impact on reproduction of the wild population and could represent the difference between decline and stability or increase of the wild population. In supporting his arguments, Miller assumed that the Condor population was roughly stable at the time, that there were normally only about 10 pairs breeding in any year, and that the birds taken would both be breeding adults, potentially disrupting two or even three active pairs and significantly decreasing wild reproduction.

In hindsight, these were extreme and questionable assumptions. As we have already seen in chapter 4, there was no good evidence for stability of the wild population at the time, and overall data strongly suggested a continuing decline. Further, there were still five confirmed pairs breeding in 1984 when the wild population had dropped to only 15 individuals. Thus, unless something fundamental had changed in the breeding tendencies of the species, it seems reasonable to believe that there may have been more than 10 pairs breeding in the early 1950s when the total population was likely 10 times as large compared to the numbers in 1984. Finally, the San Diego Zoo had clearly indicated its preferences for obtaining immature Condors rather than adults. From a modern demographic perspective, it is doubtful that the subtraction of two birds, even if they had both been breeders, from a population of roughly 150 birds (or even only 60 birds, as believed by Koford and Miller) would have had any significant impact on the survival prospects of the population. This subtraction in itself would not be expected to significantly alter the basic underlying relationship between mortality rates and reproductive rates determining health of the population as a whole.

Regardless of the merits of these considerations, the campaign of Miller and his allies carried the day, and the victory of opponents to captive breeding in the early 1950s led to a near total absence of public advocacy of this technique until the mid-1970s. By that late date, it had finally become clear, largely from progressively declining totals of birds being seen on the annual October Survey, that traditional conservation efforts for the species were not resulting in population recovery. It appeared inescapable that the species would not last much longer in the wild. Yet there had been no attempt to see if the species could at least be propagated in captivity. In fact, there was only a single California Condor in captivity at that time. This was Topatopa (pl. 74), a male that had been taken captive as a starving fledgling in 1967, then released briefly into the wild, and finally taken captive again with an injured foot. It was not until 1988 that the species would finally be bred in

Plate 74. Between 1967 and 1982, the only California Condor in captivity was Topatopa, a male retained at the Los Angeles Zoo. Topatopa first became a breeder in 1993 at the age of 26.

captivity for the first time, ironically by the San Diego Zoo, which had first proposed such efforts nearly 40 years earlier.

Reversing the Opposition to Captive Breeding

Gaining political acceptance for captive breeding of Condors was not a simple task, and captive breeding was not even proposed in the first recovery plan approved for the species in 1975, following passage of the Endangered Species Act of 1973. However, the Condor Recovery Team, an advisory group to the U.S. Fish and Wildlife Service (USFWS) that had generated the recovery plan under the chairmanship of Sanford Wilbur, became convinced shortly afterward that traditional conservation measures for the species, including the supplemental feeding program started in the early 1970s, were proving inadequate. In its 1976 recommendations to the USFWS, the team included a contingency plan that emphasized a need for captive breeding. But because of the lingering legacy of successful opposition to captive breeding, the responsible federal and state agencies were still reluctant to back this approach. Only after a detailed and gloomy analysis recommending captive breeding was provided by Jared Verner of the U.S. Forest Service in 1978, and after a special panel of the American Ornithologists' Union (AOU) and the National Audubon Society also recommended an intensive program that included captive-breeding efforts, did the federal and state agencies finally become supportive of this approach.

A crucial step in the process was the reversal of the National Audubon Society's traditional opposition to captive breeding of Condors, a reversal that was achieved largely by internal efforts of Dick Plunkett and Gene Knoder. This reversal was followed by the deliberations of the AOU-

Audubon panel under the chairmanship of Robert Ricklefs, followed by endorsement of the panel's report by the USFWS, and then by a successful effort of the National Audubon Society to lobby the U.S. Congress to create a new intensive Condor program by the end of 1979. Even though the acceptance of a need for captive breeding was intrinsic to the new program, it took several more years before this approach was actually implemented, in part because of continuing vigorous opposition from those who were still fundamentally antagonistic to captive breeding.

Establishment of a Captive Flock

The continuing intense political opposition to captive breeding encouraged responsible agencies to be cautious in how fast they allowed implementation of this technique. The first federal and state permits granted to the expanded Condor program in the early 1980s allowed the trapping of only a single unpaired female Condor to serve as a mate for Topatopa, the male already in captivity at the Los Angeles Zoo. But in part because male and female California Condors are similar in external appearance, identifying and capturing such a bird was no simple task. Further, securing a mate for Topatopa was not a particularly promising initial approach because of several problems. Topatopa was highly habituated to humans and of questionable breeding potential because of his behavioral quirks. Even if an unpaired female could be successfully identified and trapped from the wild, it was highly uncertain that this bird would prove a compatible mate for Topatopa. Also, if captive breeding were ever to have a significant role, a single captive pair would surely be inadequate to substantially aid conservation of the species either from a demographic or genetic standpoint and presumably would have to be followed by taking of additional birds of both sexes. If so, being restrictive about the sex of the first bird taken captive and

Plate 75. The first bird taken deliberately for the captive program of the 1980s was Xolxol, here being released into a large cage at the Los Angeles Zoo by Art Risser and Bill Toone of the San Diego Zoological Society. Xolxol was taken as a nestling in August 1982 when his male parent disappeared.

waiting to see how well Topatopa might breed before taking additional steps had the potential for greatly slowing the overall process and reducing chances for ultimate success when time was already very short.

As matters developed, however, the first birds deliberately taken captive in the 1980s were not taken specifically as mates for Topatopa, but were birds taken as authorized exceptions to the first program permits. The first bird captured was a nestling Condor taken in summer 1982 when its father disappeared unaccountably and its feeding rates dropped alarmingly at a nest under close observation (pl. 75). On sexing with chromosomal techniques, this bird proved to be a male, like Topatopa. In October 1982 a second immature Condor was trapped at a carcass as a potential mate for Topatopa, but when it also proved to be male, it was released again into the wild with a radio transmitter to initiate the telemetry program. The release of this bird was followed by capture of another immature as a potential mate for Topatopa in December 1982. This bird proved to be still another male, but it was retained in captivity, in part because of low body weight and an abnormal gaping problem.

Thus, early efforts to obtain a potential mate for Topatopa proved fruitless, and it was not clear how long it might take to succeed in this task. Regardless, this approach lost most of its importance and priority when it became apparent in 1982 that wild pairs might respond to deliberate replacement-clutching efforts and that a practical means might exist for establishing a captive population through that process and artificial incubation of wild-laid eggs—with only minimal impacts on the wild population. This possibility emerged as a result of detailed documentation of a case of natural replacement-clutching in the wild in 1982, possibly the most important event observed during the intensive studies of Condor breeding biology in the 1980s. In these observations, a pair lost an egg during a tussle with Common Ravens (Corvus corax) when the egg rolled out and over the lip of the nest cave, smashing on rocks below. Some 40 days later, the pair laid a replacement egg in a different nest cave.

Prior to 1982, there was no completely compelling evidence for replacement egg laying in California Condors. Carl Koford had maintained that this capacity was unlikely in the wild population because it had not been reported by historic egg collectors, although it was known to occur in many other wild avian populations. The clear documentation of this capacity in one pair of 1982 was not completely unexpected, however, as the capacity was well known by then in captive Andean Condors, both at the San Diego Zoo and the Patuxent Wildlife Research Center in Maryland (pl. 76). Also, it offered a plausible explanation for two egg layings in a single California Condor nest cave in Ventura County in 1939, as documented by Harrison and Kiff (1980). Until 1982, however, it was not politically feasible for the conservation program to propose taking of eggs as a means of establishing a captive

Plate 76. Andean Condors housed at the Patuxent Wildlife Research Center in Maryland were invaluable in developing husbandry techniques and evaluating attachments for radio transmitters before such techniques were used with California Condors.

population because of widespread fears that this would greatly harm reproduction of the wild population. After the conclusive documentation of natural replacement-clutching in 1982, the taking of eggs became a major means for establishing a captive flock, ultimately yielding nearly half of the original captives taken.

With the switch to taking eggs to form a captive flock, restrictions on the sexes of birds to be taken had to be abandoned, as it was not possible to sex eggs before hatching. By chance, nine of the first 10 eggs artificially hatched turned out to be female, producing a strong sex bias in the opposite direction from the bias that had existed with the earlier taking of three consecutive males. This bias toward females persisted throughout the rest of the period of formation of a captive flock, and it was only when the last free-flying male was trapped in 1987 that the bias was reduced to a nearly balanced captive sex ratio (with 14 females and 13 males).

Clearly, hindsight indicates that it was of little value to be greatly concerned about the sex sequence in which Condors were taken captive and pointless to assume that Topatopa would be one of the first captive breeders. The captive sex ratio wound up as close to even as could have been achieved, even though essentially all captives were taken without the decisions resting on sexes of the birds involved. Topatopa indeed proved to be an especially difficult bird to pair successfully and did not become a breeder until 1993, long after most of the other original captives had begun reproduction.

Multiple-Clutching

The efforts to induce wild Condors to lay replacement eggs proved highly successful in increasing fledgling production by the wild population and in rapidly establishing a captive flock. Between 1983 and 1986, 16 eggs were taken for artificial incubation from a total of five different pairs (including only four different females). Thirteen of the eggs produced surviving chicks, for a success rate of 81 percent, far exceeding the natural fledging success rate of 40 to 50 percent for eggs incubated in the wild. Further, four of the five pairs exhibited a capacity to lay replacement eggs, usually about a month after an egg was taken, and three pairs even laid three eggs within single breeding seasons—triple-clutching. The single female (CVF) that did not demonstrate a capacity for replacement-clutching was alive for only one egg-taking effort, and because this took place relatively late in the breeding season, whether she too might have been capable of replacement laying was not adequately tested. In all cases, pairs switched nest sites in laying replacement eggs, often moving several miles in the process, a fact that goes a long way toward explaining why replacement-clutching had not been documented historically by early egg collectors.

Once it became clear that replacement-clutching of wild pairs offered a major new hope for the conservation program, the process of taking eggs became a focal activity of the early months of each year. The process was complicated and exhausting, especially because it was essential to determine the timing of egg laying accurately to avoid the risk of taking eggs at an inappropriate stage of development. Bird eggs are vulnerable to mechanical shaking during the period when internal membranes are forming proper connections, and it was important not to remove eggs during this sensitive period (between five days and 21 days after laying in the Condor). Tracking wild pairs to accurately determine the timing of laying, however, was often a desperate scramble, as the pairs often checked many widespread nest sites in unbelievably rugged country prior to laying, and it was often impossible to predict which site might be used.

Further complications came in the actual egg-taking process. A helipad had to be constructed for each egg pickup so that the egg could be airlifted to the San Diego Zoo, and the terrain was often so rugged that it was a major job finding and clearing a site that could serve in this capacity. Weather conditions, especially wind and rain, were often marginal for helicopter flights in the back country, and it was often difficult to decide whether an egg pickup should be attempted, especially when time was running out for avoiding the sensitive period to mechanical shaking. It was also important not to flush an incubating bird from its egg in a startling manner, as the bird could accidentally damage or destroy the egg if it bolted from the nest cave in a panic. This problem was averted by very slow, but not silent, approaches to nest caves that elicited curiosity rather than panic from the incubating birds. Eggs were handled only with sterile gloves and were quickly placed in incubator suitcases that brought them back to normal incubation temperatures for the several-hour helicopter trips to the San Diego Zoo, where they were placed in commercial incubators (pls. 77, 78).

As mentioned above, only three eggs taken in the 1980s failed to produce surviving young under artificial incubation.

Plate 77. Inside a Condor nest of 1984, Rob Ramey places a California Condor egg in a carrying case for transport to the San Diego Zoo for artificial incubation.

Plate 78. A helicopter lands to pick up a Condor egg and transport it to the San Diego Zoo in 1984. Personnel of Aspen Helicopters of Oxnard provided transport for all the eggs taken during the years of egg removals.

One failure involved a very early embryonic death, another involved a chick that died shortly after hatching due to a yolk-sac infection, and the third involved a physically abnormal and inviable chick with only a partial skull and deformed limbs. None of these failures had any clear relationship with the egg-removal and artificial-incubation procedures, and it seems likely that all three eggs would also have failed if left in the wild. Fortunately, none of the egg failures occurred during the first year of egg taking, which might have caused significant political resistance to further replacement-clutching efforts.

The total production of young in 1983, including young hatched in the wild, was six individuals, a major increase above the average of two young per year that had been produced by the wild population in the previous three years. Results were even better in 1984, when seven young were produced. But by 1985, the number of egg-laying pairs in the wild had crashed to a single pair. Although this pair was successfully triple-clutched, producing two surviving young, the brief era of massive increases in production by the wild population was over, not because of any effects of the replacement-clutching operations, but because of a simple dearth of breeding birds.

The single egg-laying pair of 1985 lost its female to lead poisoning at the end of the year, but in 1986 a new pair formed among the five surviving birds in the wild and laid two eggs, one of which produced a fledgling in captivity. These were the last eggs produced by the historic wild population.

An Early-Release Proposal and the Crisis of 1985

With the outstanding success of multiple-clutching efforts in 1983 and 1984 and the rapid progress in assembling a captive flock, the Condor Recovery Team developed a comprehensive plan that would not only continue the accumulation of captive progeny from the five wild breeding pairs then in existence but would also begin to bolster the wild population with some of the young produced by multiple-clutching. Specifically, this plan, as finalized and approved by all involved agencies in 1984, proposed a goal of an initial captive flock of 32 individuals and that once five progeny were obtained for the captive flock from any pair, further progeny from the pair would be channeled into an early-release program into the wild. As two pairs were already represented by five young apiece in the captive flock in 1984, it appeared feasible to begin releases of further progeny from these pairs in 1985, assuming the pairs survived to lay eggs in that year.

Unfortunately, neither of these pairs survived until the breeding season of 1985, and during the winter of 1984–85 the wild population lost four of the five pairs active in 1984. Thus, not only were there no young Condors available to commence releases under the terms of the Condor Recovery

Team plan, but the entire logic of the plan had been compromised by the extremely heavy mortality of breeding pairs. A full 40 percent of the wild population of Condors had disappeared in just a few months and nearly all reproductive potential of the population had vanished with them, whereas the captive population of 16 individuals still remained far below the level identified as a goal by the Condor Recovery Team. There was no reasonable expectation that either the goal of establishing a genetically adequate captive population or the goal of conducting meaningful releases could be met on the basis of multiple-clutching of the single remaining wild pair.

Because of these discouraging facts, some participants in the program, including ourselves, argued that the most conservative strategy to follow was to fully recognize the basic inviability of the wild population, to abandon the idea of near-term releases into the wild, and to attempt to get as close to a genetically viable captive population as might be possible by trapping in the remaining nine wild birds. Under this thinking, release efforts into the wild should not be allowed to compromise the chances of achieving a viable captive population and should not be attempted until that goal was achieved and important limiting factors were effectively addressed in the wild—factors that were clearly much more menacing than had been recognized earlier.

Others recommended leaving most of the surviving wild birds in the wild and going ahead with releases of several captives in the near term in an effort to sustain the wild population as long as possible and to ensure continued protection of important Condor habitats. Many espousing these recommendations were also clearly concerned about the possibility that taking the last wild birds captive would mean the end of research and conservation program as it then existed. Nevertheless, the U.S. Forest Service immediately pledged that it would continue to protect all its Condor sanctuaries so long as a prospect for reestablishment of captives into the wild existed. In addition, many of the people urging capture of the last birds argued that the research and conservation program could be sustained by a focus on temporary experimental releases of surrogate Andean Condors during the period all California Condors were in captivity (a solution to program-protection concerns that was indeed eventually implemented).

The position favoring capture of all the remaining wild birds was quickly adopted by the CFGC, whereas the position favoring the leaving of most birds in the wild and going ahead with early releases became the initial position of the USFWS and National Audubon Society. Because both the CFGC and the USFWS had to authorize any actions in the program, however, the only point in common between the two agencies during early 1985 was that both favored trapping of three of the remaining nine wild birds into captivity. These three birds were captured during summer 1985. Meanwhile, the deadlock on the issues of releases and capture of the remaining six wild birds continued until almost the end of 1985.

Once the dimensions of the crisis became clear in spring 1985, the Condor Recovery Team immediately recommended that there should be no near-term releases of captives into the wild, but the team was initially split on the issue of capturing the remaining wild birds. By late summer 1985, however, the Condor Recovery Team achieved a consensus that at least three of the six birds still in the wild should be captured, and by November 1985, the team was unanimous in recommending that all the remaining wild birds should be captured. This also became the position of the USFWS in December, following a determination that one of the six birds still in the wild was suffering from acute lead poisoning, the stress that had by then emerged as probably an overwhelming threat to the wild population.

Despite the consensus that had been achieved among the CFGC, the USFWS, and the Condor Recovery Team by December 1985, actual trapping of the last wild birds took more than an additional year. The process was first delayed by an unsuccessful lawsuit to prevent their capture mounted by the National Audubon Society and later by protests of a group of Native Americans on cultural and religious grounds. The last bird was not captured until Easter Sunday 1987.

Reproductive Performance of the Captive Flock

The 27 birds in the original captive flock of 1987 included seven birds captured as adults, seven birds taken as nestlings or immatures, and 13 birds taken as eggs. Ten of these birds were adult in 1987, and two of these adults at the San Diego Wild Animal Park produced the first captive-reared nestling in 1988 (pl. 79). By the following year there were three fertile pairs that produced a total of four captive fledglings at the San Diego Wild Animal Park and the Los Angeles Zoo. Produc-

Plate 79. The first breeding of California Condors in captivity involved this pair at the San Diego Wild Animal Park in 1988. Both birds were trapped as free-flying adults in the last years of the historic wild population.

tion totals at both institutions increased steadily in the early 1990s, aided by routine multiple-clutching of captive pairs; by 1996 the Peregrine Fund in Boise, Idaho, became a third institution successfully breeding the species.

By summer 1998, only 10 years after the first breeding in captivity, there were more than 150 Condors in existence, a total matching the estimated number of Condors alive in 1950. Moreover, by the following year, all 27 of the original captives had bred successfully and there had been only a single mortality among the original captives, a record of achievement rarely matched in captive breeding programs. Clearly the captive population was quickly proving demographically vigorous, and early fears that the species might not breed readily in confinement were fast becoming worries of the past. From a population low point of 22 individuals (including both captives and birds in the wild) in fall 1982, the total numbers of Condors had rebounded with a rapidity that few had anticipated or thought possible.

Captive Breeding and Rearing Procedures

Once captive birds were sexed by chromosomal procedures, they were separated into pairs whose members were believed to represent different family lines, a strategy aimed at avoiding pairings of closely related birds. Each pair was housed in a large flight cage containing a spacious wooden nest cave, a drinking and bathing pool, and appropriate above-ground perches (pl. 80). Birds were kept visually isolated from human keepers and fed a variety of carrion sources. Closed-circuit TV cameras allowed remote observations of birds and nest chambers without disturbance and quick detection of any eggs laid. Eggs were carefully checked for fertility by candling, and in the early stages of the program chicks were helped out

Plate 80. Off-exhibit breeding cages for Condors at the San Diego Wild Animal Park allow the birds considerable space for flight and exercise on foot, as do the cages at other facilities.

by human keepers during the hatching process to minimize energy loss in the chicks as they entered the nestling phase (pls. 81–83).

From the very start, efforts were made to maximize captive reproduction by encouraging replacement laying through removal of first-laid eggs for artificial incubation. Artificially incubated eggs, however, produced chicks that had to be

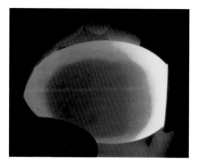

Plate 81. Checking a newly laid Condor egg for fertility involves a process called candling, which involves shining a bright light through the eggshell. An embryo first becomes visible at about four to five days of age.

Plate 82. To ensure successful hatching, many of the first captive Condor chicks were carefully helped out of their shells.

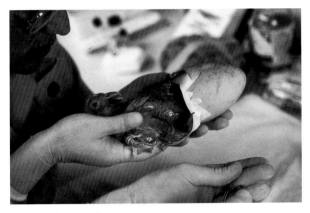

Plate 83. Cyndi Kuehler assists the hatching of Sespe, a chick that barely survived excessive incubation neglect from the CC pair while still in the wild.

reared artificially by humans. To reduce the chances that such chicks might wind up psychologically imprinted on humans, they were routinely fed at an early age by hand puppets resembling adult Condors that were operated by human caretakers from behind visual barriers (pl. 84). Later in the

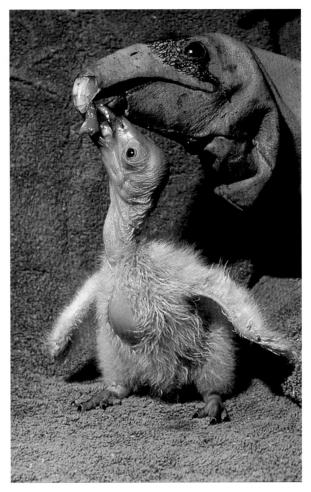

Plate 84. Because of a reliance on multiple-clutching in captivity, most captive California Condor nestlings have been puppet-reared, rather than parent-reared. Puppet-reared Condors have proved successful breeders in captivity but have shown problems with excessive human orientation in releases to the wild.

nestling period, when these chicks became capable of feeding themselves, they were commonly placed together in cages to allow socialization with other members of their own species. As the size of the captive flock began to increase significantly, however, parent Condors were increasingly encouraged to rear their own chicks—usually from their second-produced eggs of the season. Efforts have often been made to allow chicks from first-laid eggs to spend part of their nestling periods in the company of mentor adults, rather than simply in the presence of other chicks, although these mentor adults have not assumed the roles of true parents and have not actually fed and cared for the chicks.

The various rearing procedures used have clearly been successful in producing birds that have been capable of further captive reproduction, although as Hartt et al. (1994) showed, chicks reared together have shown significant difficulties in forming successful pair bonds with their early cage mates in later life. Further, as we shall see in the next chapter, artificially reared Condors have generally shown behavioral problems, such as excessive tameness and attractions to human structures, when released into the wild.

Genetic Concerns

The original captive flock of 27 individuals was known to include many individuals that were closely related to one another as siblings or as parents and their offspring. But how closely related to one another many of the birds were, was still an unknown. Nuclear DNA analyses by Geyer et al. in 1993 suggested that the Condors still alive all clustered into three clans that were relatively closely related to one another. Further, in comparison to Andean Condors tested from a variety of captive sources, the surviving California Condors appeared to be significantly depleted in genetic diversity. Additional genetic studies by Chemnick et al. (2000) and Adams (2002) revealed only a very limited number of maternal lines of mito-chondrial DNA in the surviving birds. One of these, found only in Topatopa, will disappear from the population when he dies, as he is male and cannot transmit it to his progeny.

Thus, there are reasons for concern as to whether the cap-tive flock might suffer from genetic difficulties. Although the quantitative production of young by the captive flock has been gratifyingly substantial, there have been a number of problems to emerge that may reflect genetic stress. These problems mainly concern deficiencies in egg fertility, malpo-sitions of embryos, and congenital abnormalities.

Overall fertility of eggs laid in captivity through 1998, at 75.9 percent, was considerably lower than the fertility of eggs that had been documented in the wild (100 percent) and was lower than levels normally seen in other wild bird popula-tions (usually 80 to 90 percent or higher). Nevertheless, most of the infertile eggs in captivity have been attributable to inex-perience and behavioral incompatibilities within some pairs —the latter reflected in failures to perform adequate copula-tions. Because Condors have been routinely force-paired (for genetic reasons), rather than allowed to freely choose among potential mates, one might expect to find mate compatibility problems to be relatively frequent. Fertility within pairs has showed a strong tendency to increase with time and to be much higher in pairs with some previous breeding experi-ence—additional factors suggesting that the primary causes of infertility have been behavioral, rather than genetic. For pairs producing chronically infertile eggs, the most efficient remedy has proved to be re-pairing the birds involved with new and more compatible mates.

Malpositioning of embryos within eggs affected 11 per-cent of the fertile California Condor eggs laid in captivity through 1998, and although not all these cases resulted in failed hatchings, this rate appears to be elevated in compari-son with the rates seen in Andean Condor eggs, about 5 per-cent, and eggs of various strains of Domestic Chickens (Gal-

lus domesticus), usually about 8 to 9 percent. Malpositioning can be caused by both environmental and genetic factors, so the apparently elevated incidence in California Condors could trace to genetic factors, although this is not certain.

More clearly associated with genetic causes have been several cases of congenital deformities (chondrodystrophy) that have appeared in progeny of one pair of captive Condors and in a sibling of one of these pair members. Katherine Ralls and her colleagues (2000) have suggested that although both members of the pair in question appear normal, they are both likely carriers of the gene causing the malady and that the malady only develops in progeny receiving the gene from both parents—that is, progeny that receive two doses of the gene. Because both members of this pair have many close relatives in the captive flock, this gene may be present undetected in many other carriers in the flock. Without an efficient means to detect the presence of this gene in unaffected carrier individuals, deliberate attempts to eliminate this gene from the captive flock would entail preventing breeding by a large fraction of the existing captives.

In addition to the preceding difficulties, one female in the flock has several times produced eggs with abnormal air-cell formation. This is a malady that could have a genetic cause because it has only appeared in this one bird, although not necessarily.

How serious the potential genetic difficulties may prove to be depends in part on whether the expression of these and other difficulties may be more detrimental in birds released into the wild than in birds protected by a benign captive environment. Whether Condors returned to the wild may be significantly handicapped by genetic problems and whether this could undermine the ultimate success of releases is not yet obvious in release results, but this does not yet offer a clear resolution of this question.

At this point, it is also unclear that very much could be done to deliberately alleviate genetic problems. Attempting to remove individuals carrying genetic defects from the captive gene pool poses as yet unresolved difficulties in identifying such individuals and risks losing other genetic characteristics possessed by these individuals that might be beneficial. No consensus has been reached that this might be a wise approach and no such an approach has been implemented. Instead, the breeding program has proceeded ahead mainly with a strategy of avoiding pairings within clans and encouraging all individuals to breed, relying on a belief that genetic problems will prove to be sufficiently minor that they will not sabotage the ultimate success of species recovery. If the species can successfully survive whatever genetic stresses it presently faces, there is every reason to hope that it may eventually reaccumulate greater and more usual levels of genetic variability through normal gene-mutation processes and that natural selection will eventually eliminate deleterious genes, or at least reduce them to very low frequencies in wild populations.

IN THE VIEW OF most conservationists, the primary goal of efforts on behalf of the California Condor has always been the maintenance or re-creation of wild populations that are self-sustaining and behave in a species-typical way. Most also feel that although captive populations can be an important means to that end, they should not be considered an ultimate conservation goal in themselves. One fundamental reason for this is that real-world captive populations do not stop evolving, even when comprehensive efforts are made to slow down this process. They ultimately change genetically and behaviorally into populations that are best adapted to captivity and have characteristics that are significantly different from those of wild populations. As such, they become progressively more difficult to reestablish in the wild with time. This is not to say that long-term captive populations might not have their own intrinsic worth for exhibit and other purposes, but that their values for wildlife conservation tend to steadily decline over time. The "ark" philosophy of preserving species in captivity over the long term is fundamentally flawed by this problem.

Maladaptive changes also occur in the short term in captivity. With higher vertebrate species, it is often far easier to create captive populations from wild populations than the reverse, because many of the adaptations that allow such animals to function successfully in the wild are complex learned behaviors that are commonly lost in the first generation of captive offspring. Examples include such capacities as knowing where to forage at various seasons and knowing how to respond properly to various potential predators or competitors, including humans. It can often be very difficult to re-create properly adaptive behaviors in releases of naive animals in the absence of experienced wild members of the species capable of acting as role models for the needed behaviors. For this reason, it is usually wiser for a conservation program to continuously maintain wild populations of a species into which captive-reared stock can later be introduced, than to allow all wild populations to disappear completely and then attempt to re-create them from scratch.

Unfortunately, the conservation program for the California Condor began formation of a captive flock only as the species was about to be lost, and because the wild population crashed so precipitously toward the end, a genetically viable captive flock could only be approached by removing all the very last birds from the wild. This strategy, however, precluded the option of maintaining a wild population to serve as a behavioral template for later releases of captive-reared stock, except insofar as some of the original birds taken captive might survive long enough to be released together with succeeding generations in reestablishment efforts.

As release strategies were being developed for the Condor, however, there were two reasons why it was actually unwise to release wild-experienced birds together with naive captive-reared birds in the first releases. One reason was the need to produce further captive progeny of founder wild-experienced birds to maximize genetic diversity of the captive flock. The other reason was the nature of the primary limiting factor faced by the species in the wild and the fact that this limiting factor was not corrected by the time of first releases, as is discussed below.

A Rationale for Releases

The overriding principle of animal reintroductions is that reestablishment efforts should await the correction of the limiting factors causing original extirpation from the wild. Otherwise there can be no realistic hope of achieving self-sustaining wild populations. In the case of the California Condor, the principal limiting factor identified in research of the 1980s and in more recent years has been lead poisoning. Yet as is discussed more thoroughly in the next chapter, there was no promising means available for quickly removing the lead threat from the wild environment in the 1980s, and the only way to justify beginning releases in the absence of such a removal was through an effort to prevent exposure of released birds to lead by attempting to keep them dependent on a subsidy of clean food. To understand how this release strategy was developed, it is valuable to first review certain aspects of the history of the conservation program in the mid-1980s.

By late 1985, it had become reasonably clear that lead contamination was an important cause of the precipitous decline of the wild population and that it was not feasible to remove the lead-poisoning threat in any quick manner. An attempt was made in that year to counter the lead problem by offering the remaining wild birds a steady clean-food subsidy in one part of the geographic region where the birds normally congregated in summer. But like the earlier food subsidy effort of Wilbur in the 1970s, this effort did not result in more than partial dependence of the birds on subsidy, and this was really no surprise. The last wild birds were all experienced foragers, accustomed to finding food in many regions, and foods other than those provided in the subsidy program continued to be available at the time of the feeding program, continuing to reinforce wide-ranging foraging behavior. Although the birds did sometimes take clean carcasses offered to them by the conservation program, their entire behavior was geared to wide prospecting for food, and they had no previous experience with a food supply that was limited to a single location.

Moreover, this was a population that was long accustomed to making seasonal shifts in foraging regions that tracked seasonal changes in food availability, and indeed the birds followed their usual patterns in fall 1985 when they mostly moved to a region far from the summer subsidy area. Attempts were made to move the subsidy to the location where birds were concentrated, but success in attracting birds to feeding sites in this second location was modest. When one of the last birds came down with lead poisoning in the fall, it became clear that the subsidy program simply was not effectively controlling the feeding behavior of experienced wild birds. This instance of poisoning was the crucial factor leading to a decision by the U.S. Fish and Wildlife Service (USFWS) to trap all the last birds into captivity.

Yet the failure of the feeding program to control the foraging behavior of experienced wild birds still left open the possibility that birds that had not yet learned wide-ranging foraging behavior might be trainable to limit their food intake to a geographically confined clean food supply. In particular, first releases of captive-bred California Condors into the wild were proposed on the possibility that naive released birds might be safely confinable to a controlled food supply in a remote location lacking abundant alternative foods and thus might remain unexposed to existing lead-contamination threats. Under this sort of scheme, to be sure, it would not be wise to release experienced birds from the historic wild population together with naive captive-reared birds, as the former could be expected to quickly return to their wide-ranging foraging habits and had the potential to lead captive-reared birds into the same behavior, exposing all birds to lead contamination.

The hope was that while first releases were being conducted, there would also be an effort to truly remove the lead contamination problem, either by some sort of legal restrictions on use of lead ammunitions or by acquisition of enough habitat that could be maintained free of lead contamination to sustain a population. With success in such efforts it might ultimately become possible to get birds off controlled food subsidy into a natural foraging situation.

A possibility existed, of course, that naive captive-reared birds would not remain conservative in their movements or closely dependent on subsidy after release. If that occurred, the whole justification for continuing releases would come into question, unless effective efforts might be initiated to actually remove the lead contamination threat from the wild environment.

First Releases of Andean and California Condors

To test basic release techniques and the feasibility of keeping birds dependent on a clean food subsidy in a relatively remote region, temporary experimental releases of surrogate Andean Condors *(Vultur gryphus)* were conducted in the Sespe Sanctuary vicinity, starting in 1988 (pl. 85). The birds released in these efforts were puppet-reared individuals raised at the Los Angeles Zoo, and release procedures largely followed those that had been used in an experimental release program of Andean Condors into the wild in Peru during the early 1980s (pl. 86). Those earlier Peruvian releases, which had been run by Mike Wallace (Wallace and Temple 1987b), had the advantage of introducing birds into a natural wild population from which naive released birds could learn appropriate behavior patterns. The results with these earlier releases were very en-

Plate 85. Pens for experimental releases of captive-reared fledgling Andean Condors in the late 1980s were mostly sited in the Sespe Sanctuary. Birds in these releases remained largely dependent on food subsidy but steadily expanded their movements and frequented developed areas without apparent hesitation.

Plate 86. Temporary experimental releases of Andean Condors in California were limited to juvenile females, which closely resemble California Condors in size and habits.

couraging, with good survival of the released birds and apparent full integration of these birds into the wild population within a very few years.

Unfortunately, Condor releases in the United States have all lacked the same advantage of potential integration of released birds into existing wild populations of properly behaving Condors, and many of the problems experienced in these releases have apparently traced to this difference. These releases (map 5), which were also supervised by Mike Wallace in the early years, have been carried on by a variety of organizations in later years (the USFWS in southern California, the Ventana Wilderness Society in the region near Monterey, the Peregrine Fund in the Grand Canyon region of Arizona, and most recently the Mexican government in northern Baja California).

Early survival of the first birds released in the United States—surrogate Andean Condors released in southern

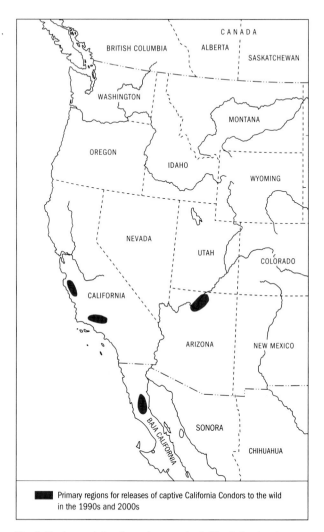

Map 5. The four main locations where captive California Condors have been released into the wild since the early 1990s include regions in California, Arizona, and Baja California, although birds are now ranging far beyond the boundaries of the release areas. All release areas fall within the known historic range of the species. Solid lines represent coasts and rivers.

California—was reasonably good, and the birds initially stayed nearly completely dependent on clean-food subsidy, with no immediate problems developing with lead poisoning, although the movements of the birds tended to steadily expand. These early results were encouraging enough that releases of puppet-reared California Condors followed the experimental Andean Condor releases, starting in early 1992 (pls. 87, 88). The Andean Condors were all returned to captivity by later in 1992.

The emphasis on puppet-reared birds, rather than parent-reared birds, in early releases was a result of the policy of routine multiple-clutching of captive pairs that was being followed to increase the size of the captive flock as fast as possible. With near maximal multiple-clutching efforts in place, the majority of progeny produced cannot be parent-reared. However, despite conscientious efforts to avoid the exposure of chicks to humans through puppet-rearing and isolation of the birds from human contact prior to release, the first released

Plate 87. The first release of two captive-reared California Condors was a mixed release with two Andean Condors in January 1992. Like the Andean Condors released earlier, all these birds showed tendencies to investigate civilized areas.

Plate 88. Released California Condors, like released Andean Condors, have initially been maintained on clean food supplies provided near the sites of release but in time have begun to take food from other sources.

birds of both species soon began to develop problems with excessive attractions to human structures and with strong tendencies of the birds to approach people without evident fear. Such behavioral problems have continued intermittently to the present.

The range of misbehaviors seen in released birds has been wide and has included incidents such as Condors landing on automobiles to destroy windshield wipers and weather stripping, Condors entering campgrounds to rip up tents and sleeping bags, Condors chasing people down back-country trails, Condors habitually landing on human structures such as oil-well pads and buildings, and even Condors entering the bedroom of a resident of Pine Mountain Club to rip up his mattress and unoccupied shorts (pl. 89). One bird was reported seen walking around with a loaded .38-caliber revolver in its bill, gripped by the trigger mechanism—apparently an item pulled from a camper's backpack. Such behaviors were simply unknown for the historic wild population. The eagerness of the birds to invade civilized regions was the apparent

Plate 89. In 1994, several released California Condors repeatedly invaded the village of Pine Mountain Club, where they vandalized structures and received food from residents. Birds from this release returned with additional Condors to again vandalize this community in 1999.

underlying cause of several cases of birds being lost to collisions with overhead wires in the early releases.

A Conference on Behavioral Problems

In an effort to combat the behavioral problems being seen in releases, the USFWS convened a workshop of outside experts in 1994. Two principal approaches were recommended by the attendees: (1) various forms of aversive conditioning to teach birds to avoid power poles and humans and (2) releases of birds raised by their own parents in naturalistic field enclosures. For the naturalistic releases, birds would never encounter typical rectangular human structures during development and would be completely isolated from visual and auditory contact with humans and human artifacts other than cage wire, both before and after release.

No thoroughly naturalistic releases have yet been carried out, but releases immediately subsequent to the workshop involved moves of release facilities to more remote locations and employed a variety of aversive-conditioning techniques. A promising reduction in apparent attractions of Condors to utility poles was achieved by training of the birds with dummy utility poles equipped with electroshock wiring (pls. 90, 91). This training was also associated with a reduction in frequency of collisions of birds with power and telephone lines.

Less success has been achieved in reducing the willingness of released Condors to approach humans and human structures. In spite of a variety of aversive-conditioning efforts, such as repeatedly capturing prerelease birds with nets to attempt to make them fear humans, human-oriented behavior still continues erratically. In part, the problem of attraction to humans in released Condors has been the apparent result of bystanders sometimes offering food to birds that have closely approached them, a problem reminiscent of the his-

Plate 90. Early-released Condors, including this Andean Condor, showed a strong tendency to land on power and telephone poles, and a number of birds were lost to collisions and electrocutions in early releases. Historic wild Condors were not known to use utility poles as perches.

Plate 91. Apparent success in discouraging released birds from landing on utility poles has been achieved by prerelease and postrelease training with dummy utility poles wired with electroshock mechanisms, shown being installed here at a Santa Barbara County release site.

toric troubles with overly tame and aggressive bears in Yellowstone National Park.

Perhaps the most encouraging of the early releases were those conducted by the Ventana Wilderness Society in the late 1990s that involved birds reared by their own parents, although in a zoo environment. These releases followed earlier releases of puppet-reared birds in the same location that had been aborted because of excessively human-oriented behavior of the released birds. The parent-reared birds were distinctly wilder than puppet-reared birds prior to release, and they showed much less tendency to approach humans and human structures and vehicles after release. Unfortunately, this group of birds fairly soon came in contact with a group of puppet-reared birds from another release and joined that group, adopting many of the human-oriented behaviors of that group. There was no opportunity to see if their initial good behavior might have lasted if the parent-reared birds had remained isolated from other Condors.

In recent years the zoos have been making great efforts to increase the production of parent-reared Condors for release. They have also been making increasing efforts to place prerelease puppet-reared Condors together with mentor adults, instead of just with other juveniles. Even though many of the recently released birds have been parent-reared, they have been introduced together with puppet-and-mentor-reared birds into populations containing puppet-and-mentor-reared birds from earlier releases, and there have not yet been any follow-up releases of parent-reared birds isolated from puppet-and-mentor-reared birds.

Additional Problems

Perhaps the most discouraging aspect of early releases has been the resurgence in cases of lead poisoning both in California and Arizona, reflecting an increased tendency for birds to feed on carcasses outside the subsidy programs. Such instances began to crop up in 1997, and since then, there have been lead-caused mortalities of at least five birds and more than 30 acute lead poisonings necessitating emergency chelation therapy. As analyzed by Meretsky et al. (2000, 2001), the mortality rates being seen in releases were much too high for population viability, and much of the problem was due to lead poisoning. Clearly the subsidy program, as practiced in early releases, was not providing satisfactory long-term control of the lead-poisoning threat, just as it failed to rescue the wild population in 1985. Efforts to keep the birds on clean subsidy have been redoubled since lead poisoning incidents developed in the late 1990s, but it remains uncertain how successful these efforts may prove in the long run. In any event, populations maintained on food subsidy are not in the last analysis self-sufficient, and ultimately there is a need for true removal of lead-contaminated carcasses from release environments if fully self-sustaining wild populations are ever to be achieved.

Meanwhile, starting in 2001, some of the released birds have begun breeding attempts in the wild, both in California and in Arizona. But although 13 eggs were laid in the wild by 2003, only one of these yielded a surviving fledgling. Results of the 2004 breeding season so far have been somewhat better, with three pairs currently raising nestlings.

Causes of the frequent failures in early breeding attempts are only partly known, and it seems possible that some failures may simply have been due to inexperience in first-time breeders. Certain failures, however, raise concerns about potential underlying behavioral problems in the release populations. For example, two early failed breeding attempts involved groups larger than pairs (a trio and a quartet) occupying single nest caves—something never observed in the historic wild population. Also raising concerns have been fatal and near-fatal cases of nestling Condors ingesting large quantities of trash (such as fragments of glass, plastic, and metal objects), causing impaction of their digestive tracts. These cases may well be traced to a continuing tendency of parent birds to frequent human-disturbed sites such as oil well pads (where such trash is abundant), a tendency unknown in the recent historic wild population.

In addition, certain of the first egg-laying pairs of released Condors have so far proved only erratic in attempting to breed, and the number of breeding pairs still remains relatively low. Egg-laying pairs in the historic wild population were very regular in breeding, and most adults were involved in breeding. It seems possible that the release populations may be having some difficulties in forming normal pair bonds between potential breeders.

Thus, although it is encouraging to see release populations initiating reproduction, and encouraging to see signs that breeding success may be improving in the most recent years, there have been at the same time some significant problems in breeding performance, and it is not yet clear how much difficulty such problems may represent in the years ahead.

Summary of Progress

The reintroduction program to date, which now involves releases in California, Arizona, and Baja California, has placed many California Condors back into the wild, in substantial part maintained on controlled food supplies of carcasses free of lead contamination. These populations have shown many positive characteristics, including a tendency to settle in historic nesting areas such as the Sespe Sanctuary and a willingness to initiate breeding activities. For those keenly concerned about the future prospects of the species, especially those who have labored long and hard in conservation efforts on behalf of the Condor, these results are a most welcome indication that the species very likely can be successfully reestablished in the wild from captivity, and that the whole captive approach is not just another blind alley in the long fight to recover the species.

Nevertheless, the goal of demographically self-sustaining wild Condor populations free of abnormal human-oriented behavior still lies ahead, and a number of problems still await resolution before this goal is achieved. Despite the provision of clean-carcass subsidy in all releases, lead poisoning has unfortunately reappeared as a major mortality threat, and it appears unlikely that self-sustaining wild populations can be achieved in the absence of a real resolution of this threat. In addition, released populations have exhibited chronic problems with abnormal attractions to humans and human structures. Although these problems have been episodic and variable, they are still being seen, and it is not clear that they will fully disappear spontaneously.

Meanwhile, although more than a dozen breeding attempts have been seen in released Condors, the pairs laying eggs so far have not all been consistent in attempting breeding, and their overall success in producing fledglings has been low. Whether breeding effort and success may rise to historic levels without further improvements in release techniques is as yet unknown.

Both the excessive human orientation of released birds and their poor breeding performance may trace in part to unnatural rearing conditions in zoo environments. Although major progress has been made in increasing the amount of parent-rearing of chicks in zoo environments and there has been a shift toward placing puppet-reared chicks with mentor adults prior to release, there has been no experimentation as yet with parent-rearing of chicks in naturalistic field enclosures away from zoo environments to see if this yields better results. Potential ways to solve remaining problems, many of them recommended by people both inside and outside the conservation program, form the primary subject of the final chapter.

THE SPEED WITH which natural environments are disappearing has accelerated enormously in recent decades, and in the absence of major changes in current trends, wildlife extinctions will surely become increasingly commonplace in the years ahead. For the near term, the survival of many endangered species has come to depend on intensive and expensive last-ditch conservation efforts. The California Condor in particular is a creature that would surely soon vanish without continuing major aid from our own species.

Unfortunately, there are legitimate concerns as to how long the resources for such efforts will continue to be available, especially with steadily increasing numbers of endangered species competing for limited conservation dollars. The funds to pursue Condor conservation efforts (currently about two million dollars annually, including both public and private sources) are substantial and could well disappear if periods of severe economic stress develop for the country. Although the Condor has surely been one of the most favored wildlife species from a fiscal standpoint in the past two decades, the more quickly it can be returned to a self-sustaining status in the wild, the more likely that it may prove to be an ultimate survivor, rather than a casualty, of future periods of stress, and the more quickly at least some of the monies invested in Condors could come to the aid of other species. These concerns argue for all possible speed in achieving full recovery of the species.

Solutions to the Lead Problem

Accumulating evidence suggests that lead contamination is a primary continuing obstacle to establishing viable wild Condor populations, and at least in early releases the provision of clean food subsidy has not proved an adequate means of side-stepping this threat. Even if food subsidy could be made completely successful in countering lead problems, most observers believe that it is not a satisfactory or desirable long-term conservation commitment. For real success in establishing truly independent and self-sustaining wild populations, there is no apparent alternative to actually removing lead contamination from the wild.

How to remove lead contamination from the Condor's food supply has been long debated. Seemingly, the problem could be solved quite directly by banning all hunting and other shooting of wildlife species in Condor range. In practical terms, however, such a ban would doubtless generate major opposition from hunters and other shooters and could well result in deliberate retaliation against released birds. Further, in view of traditions of widespread poaching in Condor range, as well as elsewhere, it is not at all clear that a hunting ban could be effectively enforced. Because of these major problems, responsible agencies have never proposed trying this solution.

A less contentious solution would be to purchase the historic foraging range of the species and convert these lands into no-hunting preserves. The monies needed for this solution, however, are far beyond the resources of the Condor program. Although some significant portions of the foraging range have recently been purchased and put into reserve status by various organizations, as described in chapter 7, these lands make up only a modest fraction of the full foraging range of the last population.

Still another means to solve the lead contamination problem has recently appeared in the form of various alternative nontoxic ammunitions for hunting. Most current hopes for success currently reside in this approach. A number of types of new ammunitions have been developed that appear, at least from preliminary testing, to be fully nontoxic when ingested by wildlife. Moreover, some of these ammunitions, especially those based on composite mixtures of tin, tungsten, and bismuth (TTB ammunitions), appear to be closely comparable, and even superior in some respects, to lead in their hunting characteristics. A conversion to these ammunitions would not threaten anyone's abilities to pursue traditional hunting or shooting activities and would potentially benefit many species beyond Condors, assuming the new ammunitions prove as fully nontoxic as tests so far indicate.

To fully remove the lead threat, the new nontoxic ammunitions would have to completely replace lead ammunitions. A precedent for such a replacement exists in the switch from lead shot to nontoxic steel shot for waterfowl hunting on federal lands in the 1980s. Although this transition was politically turbulent, it was eventually accomplished, even though steel shot was distinctly inferior to lead shot in its hunting characteristics. Because the new nontoxic ammunitions do not pose the same problems of inferior hunting characteristics that were posed by steel shot, one might conclude that a full phaseout of lead ammunitions could now be an easier process. In particular, the new TTB-based ammunitions (both shot and bullets) have similar densities and ranges to lead ammunitions and are soft enough to be free of the threats of damage to gun barrels that were initially posed by steel shot. And although steel bullets do not represent a satisfactory alternative to lead bullets because of major deficiencies in killing power, the new TTB-based ammunitions are comparable to lead in killing power.

The main obstacles to adoption of new nontoxic ammunitions are not the hunting characteristics of the ammunitions but the fact that these ammunitions are not yet fully available commercially and cannot be fabricated as cheaply as lead ammunitions. Nevertheless, the initial prospective price differentials are not huge (perhaps only on the order of 10 to 15 percent) and will likely shrink with improved manufacturing technologies and with cost reductions derived from increasing volumes of sales. The U.S. military has recently announced its intentions to convert fully to such ammunitions in a "green bullet" program, and if it follows through with these intentions, such a conversion could greatly benefit both public acceptance of these ammunitions and the cost aspects of their manufacture.

In any event, ammunition costs make up only a tiny fraction of the expenses of most hunters, and it is reasonable to expect that many hunters might become enthusiastic supporters of the new nontoxic ammunitions if their advantages were effectively promoted. For these users, the conservation and human health benefits associated with the nontoxic ammunitions could be expected to outweigh purchase-cost disadvantages.

In fact, it may prove wise to promote the phaseout of lead ammunitions primarily on the basis of benefits to human health. The unavoidable ingestion of small amounts of lead inherent in consuming game killed with lead ammunitions is a cost that few hunters might be willing to continue to bear if they knew the full implications of this contamination. The abolition of lead in gasoline, paint, and plumbing has come about largely through recognition of the hazards to human health posed by even low concentrations of this material. Basing ammunition conversions solely on benefits to Condors would ignore these even more compelling direct benefits to human welfare, as well as the benefits to other wildlife species vulnerable to lead contamination. We know of no persuasive reasons why lead, in all its forms that can contaminate humans, should remain any more socially acceptable than asbestos.

It is often suggested that restrictions of ammunitions to nontoxic sorts need apply only to Condor release areas. However, making nontoxic ammunitions mandatory only in Condor release areas has a number of drawbacks, especially from the larger standpoint of benefits to human health and benefits to species other than Condors. Further, although a geographically limited conversion might be easier to achieve from a political standpoint than a nationwide conversion, it would likely entail greater enforcement efforts to ensure compliance than would a nationwide conversion. The compliance issue could actually be a major problem with a geographically limited ban and might preclude adequate success in reducing lead contamination of Condors. In view of credible estimates that a substantial fraction, and perhaps even a majority, of deer shot in California are poached rather than legal kills in the first place, one can only assume that significant numbers of hunters in this region might continue to use lead ammunitions in spite of a ban, assuming they could still obtain them. Only with a full nationwide conversion to nontoxic ammunitions would the ready availability of lead ammunitions ultimately disappear.

So far, federal and state governments have mostly proved reluctant to seek a mandatory phaseout of lead ammunitions, even one limited to Condor release areas. In part, this reluctance is not surprising, as governmental wildlife agencies are faced with unending crises on many fronts and are not normally anxious to increase their burdens by taking on issues they perceive as potentially controversial. To bring about an effective solution to the lead threat, it may well prove necessary for a coalition of interested outside organizations and other parties to provide appropriate pressure for change. The political problems entailed in a full mandatory conversion to nontoxic ammunitions may prove to be much less than feared by some observers if proper education efforts are made to highlight the full costs of continued use of lead ammunitions.

As of this writing, efforts to remove the lead threat have been limited to campaigns to educate hunters about the problems of lead toxicity and to urge hunters to use nontoxic ammunitions or to bury or conceal gut piles and other remains from their hunting activities. Although these efforts must clearly be considered beneficial, and may well be essential first steps, it is, nevertheless, questionable that they will reach and affect the substantial poaching community or provide an adequate overall solution to the lead problem. So long as some hunters continue to use lead ammunition and continue to fail to recover some carcasses of shot species, lead-poisoning episodes will continue. In our view, even with appropriate education efforts, leaving the use of nontoxic or lead ammunitions a voluntary choice provides only modest hopes of achieving an adequately effective solution. Some users can be expected to continue to buy the cheapest ammunitions available, regardless of health and conservation risks, and only if purchase-cost differentials between the two types of ammunition might be reversed by appropriate taxation or subsidy policies would there be reason to have much optimism about success with a voluntary system. Much more promising would be a major campaign to seek mandatory replacement of all lead ammunitions with nontoxic alternatives on a nationwide basis.

Regardless of the exact policies that might be followed, society will continue to pay the human-health and overall wildlife costs of lead contamination until real success is achieved in removing or greatly reducing all sources of lead contamination in the environment. And so long as lead ammunitions continue to be in widespread use, Condor releases presumably will continue to be plagued by demoralizing mortalities and desperate chelations of acutely poisoned birds.

Quality versus Quantity in Release Strategies

Judging from release results so far, re-creating wild populations that exhibit behavior typical of the historic population will likely require finding new ways to correct the excessive human-oriented behaviors of released birds. The release methods implemented to date, based mainly on birds artificially reared in zoo environments and various aversive-conditioning efforts, have not yet produced release populations with a close match to the behaviors typical of historic wild Condors. And even when parent-reared birds have been introduced into already established release flocks, they have generally soon adopted the behaviors of the flocks already in the wild. Although some behavioral problems have become less frequent over the years and it cannot be excluded that current wild populations may change spontaneously into fully well-behaved populations in the future, releases have now been conducted for 13 years, and some behavioral problems continue, suggesting that such problems can endure as norms likely to be adopted by all new birds introduced into the populations.

Rather than committing all hopes to the ultimate success of present approaches, we believe it would be reasonable to begin experimentation with the naturalistic releases recommended at the 1994 U.S. Fish and Wildlife Service workshop on Condor release methodology. The Condor Recovery Team endorsed such efforts in 2000, but actual implementation of this approach has not yet occurred.

Conscientious testing of naturalistic releases should be focused on releases of birds raised from hatching by their parents and housed in naturalistic field enclosures with cavelike nesting chambers that bear no resemblance to standard human structures. These enclosures should be located in release areas in remote natural habitats far from the zoo environment and should be designed to allow a complete avoidance of human interactions with the birds. In addition, to have a reasonable chance of success, naturalistic releases must be conducted in regions that are completely isolated from regions where misbehaving Condors are already present, as Condors are highly social birds that quickly learn from one another. Naive released birds, no matter what their initial quality, can be expected to quickly adopt misbehaviors if they join such birds.

In properly naturalistic releases, birds should be prevented from seeing humans or typical human structures, hearing human voices or their machines, or being handled by humans either before or immediately after release. Interactions of the birds with other Condors prior to release should be limited to interactions with their own parents, matching the natural wild situation as closely as possible both in timing and duration. In contrast to current releases, release birds should not be socialized with other release birds prior to release, because regular contact with nonparental birds does not normally occur prior to fledging in nature. In fact, breeding enclosures should be set up in isolation from one another to prevent even visual contact between prerelease birds. In addition, parent birds should be encouraged to fledge chicks from early-laid eggs in the breeding season to avoid the penalties of reduced breeding frequency entailed in fledging chicks from late-laid eggs. Release birds should be allowed first flights on a natural schedule, and provision needs to be made for them to be fed by their parents during the normal weaning period after release, perhaps through ports in release cages.

Rearing conditions such as those described above appear to offer the best chance of producing birds that are normal with respect to their avoidance of humans and human structures and are as normal as possible in their interactions with other Condors. Although there is no guarantee that such efforts would be completely successful in avoiding the behavioral problems seen so far, the present persistence of these problems in releases argues persuasively that naturalistic releases be given a conscientious try.

In the absence of naturalistic releases serving as a comparison, there is no way to establish how many behavioral problems being seen currently in releases are being caused by a lack of normal parent-chick interactions during development of many of the birds. Nevertheless, the problems seen in reproduction of released birds give grounds for concern that behavioral abnormalities may be significant enough to adversely affect the chances of achieving self-sustaining populations. Placing release birds together with one another prior to fledging and placing unrelated mentor adult Condors in cages with preflight birds are basically unnatural strategies and cannot be expected to adequately mimic natural interactions with parents. Although such strategies may well be far superior to raising chicks in complete isolation from all other Condors, they may produce unwanted later effects in behavior of the birds.

Indeed, Hartt et al. (1994) have presented evidence of later pairing problems in captive Condors placed together as nestlings, an apparent example of the phenomenon known as the Westermark Effect, a widespread behavioral mechanism to prevent interbreeding between very close relatives and perhaps an underlying cause of some of the reproductive problems seen so far in released Condors. In natural Condor populations, avoidance of breeding with early associates may function primarily in preventing Condors from pairing with their own parents at a later date, but regardless, the apparent existence of such a phenomenon argues strongly against placing birds together in nonfamilial groups prior to fledging age (especially heterosexual nonfamilial groups). And because one can indeed allow captive parents to rear their own chicks to fledging in relatively natural surroundings that do not include other Condors, this would appear to be a preferred strategy that makes the fewest assumptions about achieving the proper environment for behavioral development of chicks.

With long-lived and intelligent birds such as Condors, who develop much of their behavior through learning processes, it appears likely that the safest way to proceed in conducting releases is to approximate the natural rearing situation as closely as possible. With such species, rearing environments that preclude normal parent-chick interactions and teach chicks that rectangular wooden nest boxes are home can be expected to produce abnormal behavior in released birds, as can placement of nestlings with other nonparental birds prior to fledging. With species that are relatively hardwired in their behavior, such practices may not have as much detrimental influence as with species that are more malleable in their behavior. Condors are strikingly primatelike in their intelligence and behavioral plasticity, and the scientific literature abounds with experimental evidence of the harm of abnormal rearing environments on adult behaviors of primates.

Two main approaches have dominated attempts to reestablish wild populations of various vertebrate species from captive-bred sources: the first being an effort to mimic natural conditions in rearing procedures as closely as possible to maximize the quality of animals released and the second being an effort to maximize numbers of individuals released regardless of whether or not they have experienced the most natural rearing procedures possible. Underlying the faith in the second approach is a belief that if enough individuals with different rearing histories are released, at least some of them may survive and reproduce; if this happens, adaptive behaviors will eventually come to predominate in the release populations.

Condor releases to date have emphasized the second approach, and behavioral problems have been continuous since the start. Because Condors are very social birds and readily learn from one another, it appears there is a significant risk that some undesirable behaviors will perpetuate themselves in release populations, even when high-quality birds are introduced into these populations. It would be of great value to see what might be achieved with releases of the highest-quality birds in complete isolation from lesser-quality birds.

Much of the present continuing commitment to releases of puppet- (and mentor-) reared Condors stems from a practice of continued multiple-clutching of captive pairs, a policy that maximizes quantity rather than quality of birds produced. Multiple-clutching efforts were important and beneficial in establishing a captive flock and in rapidly expanding the initial size of the captive flock, but they have continued as a standard practice and have not been fully compatible with a focus on parent-rearing of progeny for release. With the achievement of a large captive population, the genetic and demographic needs for multiple-clutching are much less compelling now than they were, and concerns about the behavioral penalties of this approach are steadily becoming a more important consideration. If at least part of future efforts were allocated toward full testing of the naturalistic approach to releases, the chances seem reasonably good that much-improved results might be achieved. Experience in some other release programs suggests that quality of release animals can often be the most important factor in achieving success once ecological limiting factors are successfully countered.

Unforeseen Problems

Lead contamination and behavioral problems are not the only problems that might prevent full recovery of the Condor. With the present rapid pace of ecological changes, ranging from global warming and various and proliferating forms of environmental pollution to increased spread of exotic diseases, many other stresses could affect the species in the years ahead. As just one example of future uncertainties, no one yet knows whether Condors may prove susceptible to a disease that has recently invaded North America—West Nile Virus. This disease is currently causing high mortality of certain wild bird populations of the eastern U.S., and first cases are now appearing along the West Coast. Although efforts are being made to protect captive and released Condors against this threat with an experimental vaccine, it is as yet unclear how much risk this disease represents to the species and how effective any therapy might be against it. Just as human societies can never be considered invulnerable to newly emerging diseases, there is no compelling reason to assume that Condors will be safe from such threats.

When we examine the sorts of stresses that currently threaten both wildlife and human populations in general, it is clear that many of these stresses did not even exist just a few decades ago, and it is only prudent to expect that the future may likewise bring a variety of additional unforeseen problems. This is not to counsel fatalism and despair, but to urge a high level of vigilance in all conservation activities so that the chances of failure can be kept at the lowest level possible. So long as human populations and human impacts on the environment continue to expand, this will probably be the best that can be achieved.

In conclusion, the California Condor still exists, and viable wild populations of the species still appear to be a reasonable goal if the commitment and resources to continue conservation efforts with the species do not falter. Nevertheless, even if full success is achieved in countering the lead-toxicity and behavioral problems currently troubling the release program, it remains to be seen if survival and reproductive rates will become fully adequate to allow self-sustaining wild populations. Continued research into the nature of stresses faced by wild populations and into ways to counter stresses is as essential a component of future conservation efforts, as is comprehensive testing of release techniques with well-designed experiments.

The road to success in Condor conservation has always been a meandering and bumpy one, taxing the talents and endurance of dedicated program participants to the maximum. In part, this has been due to the nature of the species itself, as the Condor has always been a very difficult creature to study and understand, with a knack for doing the unexpected. The information base on the species has changed dramatically with continuing research, and this has demanded major changes in conservation strategies, especially as the species neared extinction in the 1980s. Discovery of the main causes of the species' decline has taken many decades, and research on these causes is a continuing process. Effectively countering these causes has proved to be a more challenging task than was anticipated by early workers, more because of political problems than biological ones.

Nevertheless, we remain optimistic that there is an end to the road and that it will indeed be the achievement of fully vigorous wild populations of one of our most incredible and spectacular wildlife species. Success will not likely come by ignoring problems, but by devising ways to identify and solve problems in the most effective and efficient ways possible. The existence of a demographically vigorous captive population has reduced the risks of immediate extinction enormously, and the conservation program is now within reach of full success. The remaining obstacles all appear to be solvable problems, with enough determination and skill given to the task.

Over the decades, a surprisingly large number of people, representing a great diversity of viewpoints, have contributed to efforts to conserve the California Condor. Although there have been many disagreements regarding the best strategies to follow, and progress has sometimes seemed slow, there has, nevertheless, been a tremendous amount of progress in spite of formidable obstacles. Conflicts have had a strong tendency to disappear as knowledge has increased about the nature of the problems to be surmounted.

Clearly the Condor is not a species that can be restored to vigorous wild populations just by saving some habitat or passing some protective laws. Full conservation of this species has always demanded and is going to continue to demand something more: a basic commitment of our species to the Condor's right to exist as a wild creature and a willingness to modify our own goals and behavior sufficiently to permit that existence. This will entail some sacrifices, but the rewards for success can be expected to greatly surpass the costs involved.

No one who has ever seen wild California Condors circling majestically among the clouds will ever forget the riveting nature of the experience or question why we, as a society, should make every effort possible to promote the full recovery of this species. Viable wild populations of California Condors are surely a legacy that we should strive to leave for all succeeding generations.

TIMELINE OF IMPORTANT HABITAT PROTECTION ACTIONS

1937   Establishment of the 1.9 square mile Sisquoc Condor Sanctuary in Los Padres National Forest.

1947   Sespe Condor Sanctuary of 54 square miles established in Los Padres National Forest.

1951   Sespe Condor Sanctuary enlarged to 82.8 square miles, including a prohibition on surface entry in most critical Condor areas.

1964   Creation of the 233 square mile San Rafael Wilderness Area in Los Padres National Forest.

1969   Acquisition of the 0.1 square mile San Cayetano inholdings for Sespe Condor Sanctuary by the U.S. Forest Service.

1970   Secretary of the interior places a moratorium on all oil and gas leasing in the Sespe Condor Sanctuary.

1970   Secretary of the interior takes a stand against the Sespe Water Project because of anticipated impacts on Condors.

1970   U.S. Bureau of Land Management places a moratorium on all mineral leasing activities within areas delineated as especially important to Condor survival.

1971   U.S. Forest Service prepares a habitat management plan for the Condor, setting guidelines for management of Condor habitats on National Forest lands.

1971   Acquisition of the 0.3 square mile Huff's Hole Condor area in San Luis Obispo County by the Nature Conservancy and the U.S. Forest Service.

1972	Firearms closure of Sespe Sanctuary and adjacent Condor habitat instituted by the Los Padres National Forest supervisor.
1973	California State Legislature prohibits low aircraft flights over Sespe Sanctuary.
1973	Acquisition of the 0.5 square mile Green Cabins inholding in Sespe Sanctuary by the National Audubon Society.
1975	Acquisition of the Coldwater Canyon tract inholding in Sespe Sanctuary by the California Department of Fish and Game.
1975	Acquisition of the 2.8 square mile Hopper Mountain National Wildlife Refuge adjacent to Sespe Sanctuary by the U.S. Fish and Wildlife Service as Condor feeding habitat.
1976	Designation of California Condor Critical Habitats by the secretary of the interior.
1978	Creation of the 261 square mile Ventana Wilderness Area and the 33.9 square mile Santa Lucia Wilderness Area in Los Padres National Forest.
1981	Nature Conservancy and the U.S. Forest Service complete purchases of 0.5 square mile of private parcels of Condor nesting habitat.
1981	Acquisition by the Wildlife Conservation Board of a 0.9 square mile private parcel in the Blue Ridge Condor roosting area for the Department of Fish and Game.
1983	Acquisition of the 0.2 square mile Cottrell Flat and Willett Hot Springs inholdings in the Los Padres National Forest by the U.S. Forest Service.
1983	Acquisition of 1.4 square miles in the Blue Ridge Condor roosting area by the U.S. Fish and Wildlife Service.
1983	Acquisition of 0.3 square mile of the Elkhorn Plain by the Department of Fish and Game.
1984	Acquisition of the 0.5 square mile Oak Flat and Ten Sycamore Flat properties for the Los Padres National Forest by the U.S. Forest Service.
1984	Acquisition of the 1.4 square mile Peck Ranch for the Blue Ridge Condor area by the Department of Fish and Game.

1984 Creation of the 111 square mile Dick Smith Wilderness Area in Los Padres National Forest.

1985 Acquisition of the 0.7 square mile Indian Creek parcel for the Los Padres National Forest by the U.S. Forest Service.

1985–87 Acquisition of the 21 square mile Hudson Ranch and adjacent properties to become the Bitter Creek National Wildlife Refuge by the U.S. Fish and Wildlife Service.

1992 Creation of the 343 square mile Sespe Wilderness Area, the 60 square mile Chumash Wilderness Area, the 46 square mile Matilija Wilderness Area, the 22 square mile Garcia Wilderness Area, and the 23 square mile Silver Peak Wilderness Area within the Los Padres National Forest; also expansion of the San Rafael Wilderness Area by 73 square miles and the Ventana Wilderness Area by 59 square miles.

1992 Adoption of Wild and Scenic River designation for 31.5 miles of Sespe Creek, 32.9 miles of the Sisquoc River, and 19.5 miles of the Big Sur River.

1996 Acquisition of the 148 square mile San Emigdio Ranch by the Wildlands Conservancy to create the Wind Wolves Reserve.

2001 Creation of the Carrizo Plains National Monument.

REFERENCES AND FURTHER READING

Adams, M. S. 2002. Genetic variation in the captive population of the California Condor. Master's thesis, California Polytechnic State University, San Luis Obispo.

Anthony, A. W. 1893. Birds of San Pedro Martir, Lower California. *Zoe* 4:228–33.

Bancroft, H. H. 1882. *The native races.* Vol. 3, *Myths and legends.* San Francisco, Calif.: A. L. Bancroft and Co.

Bartram, W. 1791. *Travels through North and South Carolina, Georgia, East and West Florida.* Philadelphia, Pa.: James Johnson.

Bates, C. D., J. A. Hamber, and M. J. Lee. 1993. The California Condor and California Indians. *American Indian Art Magazine* (winter):40–47.

Beebe, C. W. 1906. The California Condor. *Bulletin of the Zoological Society* (New York) 20:258–59.

Beissinger, S. R. 2002. Unresolved problems in the condor recovery program: Response to Risebrough. *Conservation Biology* 16:1158–59.

Bendire, C. 1892. Life histories of North American birds. *Smithsonian Institution Special Bulletin,* no. 1.

Bidwell, J. 1966. *Life in California before the gold discovery.* Palo Alto, Calif.: Lewis Osborne.

Bijleveld, M. 1974. *Birds of prey in Europe.* London: Macmillan Press.

Borneman, J. C. 1966. Return of a condor. *Audubon Magazine* 68:154–57.

Brown, W. M. 1993. Avian collisions with utility structures: Biological perspectives. In *Proceedings of the international workshop on avian interactions with utility structures,* ed. E. Colson and

J. Huckabee, 12-1–12-13. Palo Alto, Calif.: Electric Power Research Committee and Avian Powerline Interaction Committee.

Brown, W. M., R. C. Drewien, and E. G. Bizeau. 1987. Mortality of cranes and waterfowl from powerline collisions in the San Luis Valley, Colorado. In *Proceedings of the 1985 Crane Workshop,* ed. J. C. Lewis, 126–36. Grand Island, Neb.: Platt River Whooping Crane Habitat Maintenance Trust and U.S. Fish and Wildlife Service.

Brunetti, O. 1965. Report on the cause of death of a California Condor. Unpublished report, Sacramento, Calif.: California Department of Fish and Game.

Campbell, K. E., and L. Marcus. 1990. How big was it? Determining the size of ancient birds. *Terra* 28:32–43.

Campbell, K. E., Jr., and E. Tonni. 1980. A new genus of teratorn from the Huayquerian of Argentina (Aves: Teratornithidae). *Contributions in Science, Natural History Museum Los Angeles County* 330:59–68.

———. 1981. Preliminary observations on the paleobiology and evolution of teratorns (Aves: Teratornithidae). *Journal of Vertebrate Paleontology* 1:265–72.

———. 1983. Size and locomotion in teratorns (Aves: Teratornithidae). *Auk* 100:390–403.

Carpenter, J. W. 1982. Medical and husbandry aspects of captive Andean Condors: A model for the California Condor. In *Annual Proceedings of the American Association of Zoo Veterinarians,* 13–19.

Carrier, W. D. 1971. *Habitat management plans for the California Condor.* U.S. Forest Service.

Chemnick, L. G., A. T. Kumamoto, and O. A. Ryder. 2000. Genetic analyses in support of conservation efforts for the California Condor. *International Zoo Yearbook* 37:330–39.

Clark, A. J., and A. M. Scheuhammer. 2003. Lead poisoning in upland-foraging birds of prey in Canada. *Ecotoxicology* 12:23–30.

Collins, P. W., N. F. R. Snyder, and S. D. Emslie. 2000. Faunal remains in California Condor nest caves. *Condor* 102:222–27.

Cooper, J. G. 1870. Ornithology of California. Vol. 1., *Land birds.* Cambridge: Welch, Bigelow and Co.

———. 1890. A doomed bird. *Zoe* 1:248–49.

Cooper, J. G., and G. Suckley. 1859. The natural history of Washington Territory. New York: Bailliere Bros.

Cowles, R. B. 1958. Starving the condors? *California Fish and Game* 44(2):175–81.

Dawson, W. L. 1923. *The birds of California.* San Diego, Calif.: South Moulton Co.

Emslie, S. D. 1987. Age and diet of fossil California Condors in Grand Canyon, Arizona. *Science* 237:768–70.

———. 1988. The fossil history and phylogenetic relationships of condors (Ciconiiformes: Vulturidae) in the New World. *Journal of Vertebrate Paleontology* 8: 212–28.

Erickson, R. C., and J. W. Carpenter. 1983. Captive condor propagation and recommended release procedures. In *Vulture Biology and Management,* ed. S. R. Wilbur and J. A. Jackson, 385–99. Berkeley: University of California Press.

Finley, W. L. 1906. Life history of the California Condor I. Finding a condor's nest. *Condor* 8:135–42.

———. 1908a. Life history of the California Condor II. Historical data and range of the condor. *Condor* 10:5–10.

———. 1908b. Life history of the California Condor III. Home life of the condor. *Condor* 10:59–65.

———. 1908c. Home life of the California Condor. *Century* 75:370–80.

———. 1908d. California Condor. *Scientific American* 99:7–8.

———. 1910. Life history of the California Condor IV. The young condor in captivity. *Condor* 12:5–11.

Finley, W. L., and I. Finley. 1926. Passing of the California Condor. *Nature Magazine* 8:95–99.

Fry, D. M., and Maurer, J. R. 2003. *Assessment of lead contamination sources exposing California Condors.* Final Report to the California Department of Fish and Game.

Fry, D. M., G. Santolo, and C. R. Grau. 1986. *Final report for interagency agreement: Effects of compound 1080 on Turkey Vultures.* Department of Avian Sciences, University of California, Davis.

Fry, W. 1926. The California condor—a modern roc. *Gull* 8(5):1–3.

Geyer, C. J., O. A. Ryder, L. G. Chemnick, and E. A. Thompson. 1993. Analysis of relatedness in the California Condors from DNA fingerprints. *Molecular Biology and Evolution* 10:571–89.

Grinnell, J. 1932. Archibald Menzies, first collector of California birds. *Condor* 34:243–52.

Guthrie, D. A. 1998. Fossil vertebrates from Pleistocene terrestrial deposits on the Northern Channel Islands, southern California. In *Contributions to the geology of the Northern Channel Islands,*

Southern California, ed P. W. Wiegand, 187–92. American Association of Petroleum Geologists, Pacific Section, MP 45.

Harris, H. 1941. The annals of *Gymnogyps* to 1900. *Condor* 43:3–55.

Harrison, E. N., and L. F. Kiff. 1980. Apparent replacement clutch laid by wild California Condor. *Condor* 82:351–52.

Hartt, E. W., N. C. Harvey, A. J. Leete, and K. Preston. 1994. Effects of age at pairing on reproduction in captive California Condors *(Gymnogyps californianus)*. *Zoo Biology* 13:3–11.

Heerman, A. L. 1859. Report of explorations and surveys for a railroad route from the Mississippi River to the Pacific Ocean, 1853–56. Vol. 10, part 4, *Report upon the birds collected on the survey*, 28–80. Washington, D.C.: Beverly Tucker.

Hegdahl, P. L., K. A. Fagerstone, T. A. Gatz, J. F. Glahn, and G. H. Matsche. 1986. Hazards to wildlife associated with 1080 baiting for California ground squirrels. *Wildlife Society Bulletin* 14:11–21.

Hegdahl, P. L., T. A. Gatz, K. A. Fagerstone, J. F. Glahn, and G. H. Matschke. 1979. *Hazards to wildlife associated with 1080 baiting for California ground squirrels. Final report on interagency agreement EPA-IAG-07–0449.* Washington, D.C.: U.S. Fish and Wildlife Service, Environmental Protection Agency.

Hendron, J. 1998. Lead exposure/poisoning incidents involving California Condors. Unpublished report, U.S. Fish and Wildlife Service, Hopper Mountain NWR Complex.

Henshaw, H. W. 1876. Report on the ornithology of the portions of California visited during . . . 1875. In *Annual report upon the geographical surveys . . .* , by G. M. Wheeler 224–78. Washington, D.C.: Government Printing Office.

Hornaday, W. T. 1889. The extermination of the American bison. *Annual report of the National Museum, 1887.* Washington, D.C.: Smithsonian Institution.

———. 1911. *Popular official guide to the New York Zoological Park.* New York: New York Zoological Society.

Houston, D. C. 1974a. Food searching in griffon vultures. *East African Wildlife Journal* 12:61–77.

———. 1974b. The role of griffon vultures *Gyps africanus* and *Gyps ruppellii* as scavengers. *Journal of Zoology* (London) 172:35–46.

———. 1975. Ecological isolation of African scavenging birds. *Ardea* 63:55–64.

———. 1976. Breeding of the White-backed and Rüppell's Griffon Vultures, *Gyps africanus* and *G. rueppellii*. *Ibis* 118:14–40.

———. 1978. The effect of food quality on breeding strategy in griffon vultures (*Gyps* spp). *Journal of Zoology* (London) 186:175–84.

———. 1979. The adaptations of scavengers. In *Serengeti: Dynamics of an ecosystem*, ed. A. R. E. Sinclair and M. Norton Griffiths, 263–86. Chicago: University of Chicago Press.

———. 1980. Interrelations of African scavenging animals. *Proceedings of the Fourth Pan-African Ornithological Congress*, 307–12.

———. 1984a. A comparison of the food supply of African and South American vultures. *Proceedings of the Fifth Pan-African Ornithological Congress*, 249–62.

———. 1984b. Does the King Vulture *Sarcoramphus papa* use a sense of smell to locate food? *Ibis* 126:67–69.

———. 1985. Evolutionary ecology of Afrotropical and Neotropical vultures in forests. In *Neotropical Ornithology*, ed. M. Foster. *Ornithological Monographs*, no. 36:856–64.

———. 1986. Scavenging efficiency of Turkey Vultures in tropical forest. *Condor* 88:318–23.

———. 1987. The effect of reduced mammal numbers on *Cathartes* vultures in Neotropical forests. *Biological Conservation* 41:91–98.

———. 1988a. Competition for food between Neotropical vultures in forest. *Ibis* 130:402–17.

———. 1988b. Digestive efficiency and hunting behaviour in cats, dogs, and vultures. *Journal of Zoology* (London) 216:603–5.

———. 1996. The effect of altered environments on vultures. In *Raptors in human landscapes*, ed. D. Bird, D. Varland, and J. Negro, 327–35. London: Academic Press.

———. 2001. *Condors and vultures*. Stillwater, Minn.: Voyageur Press.

Howard, H. 1952. The prehistoric avifauna of Smith Creek Cave, Nevada, with a description of a new gigantic raptor. *Bulletin of the Southern California Academy of Sciences* 51:50–54.

———. 1962. A comparison of avian assemblages from individual pits at Rancho La Brea, California. *Contributions in Science, Natural History Museum Los Angeles County* 58:3–58.

Jackson, J. A. 1983. Nesting phenology, nest site selection, and reproductive success of Black and Turkey Vultures. In *Vulture bi-*

ology and management, ed. S. R. Wilbur and J. A. Jackson, 245–70. Berkeley: University of California Press.

Janssen, D. L., J. E. Oosterhuis, J. L Allen, M. P. Anderson, D. G. Kelts, and S. N. Wiemeyer. 1986. Lead poisoning in free-ranging California Condors. *Journal of the American Veterinary Medical Association* 155:1052–56.

Johnson, E. V. 1985. Commentary, California Condor population estimates. *Condor* 87:446–47.

Johnson, E. V., D. L. Aulman, D. A. Clendenen, G. Guliasi, L. M. Morton, P. I. Principe, and G. M. Wegener. 1983. California Condor: Activity patterns and age composition in a foraging area. *American Birds* 37:941–45.

Jurek, R. 1983. Chronology of significant events in California Condor history. *Outdoor California* 44:42.

Kahl, M. P. 1966. A contribution to the ecology and reproductive biology of the Marabou Stork *(Leptotilus crumeniferus)* in East Africa. *Journal of Zoology* 148(3):289–311.

Kalmbach, E. R. 1939. American vultures and the toxin of *Clostridium botulinum. Journal of the American Veterinary Medical Association* 94:187–91.

Kellner, A. W. A., and W. Langston, Jr. 1996. Cranial remains of *Quetzalcoatlus* (Pterosauria, Azhdarchidae) from late Cretaceous sediments of Big Bend National Park, Texas. *Journal of Vertebrate Paleontology* 16(2): 222–31.

Kiff, L. F. 1989. DDE and the California Condor *Gymnogyps californianus:* The end of a story? In *Raptors in the Modern World,* ed. B.-U. Meyburg and R. D. Chancellor, 477–80. Berlin, Germany: World Working Group for Birds of Prey.

Kiff, L. F., D. B. Peakall, and S. R. Wilbur. 1979. Recent changes in California Condor eggshells. *Condor* 81:166–72.

Knight, R. L., H. A. L. Knight, and R. J. Camp. 1993. Raven populations and land-use patterns in the Mojave Desert, California. *Wildlife Society Bulletin* 21:469–71.

Koford, C. B. 1953. The California Condor. *National Audubon Society Research Report,* no. 4:1–154.

———. 1979. California Condors, forever free? *Audubon Imprint* 3(9):1–3, 6–7. Santa Monica Audubon Society.

Kuehler, C. M., D. J. Sterner, D. S. Jones, R. L. Usnik, and S. Kasielke. 1991. Report on captive hatches of California Condors *(Gymnogyps californianus):* 1983–1990. *Zoo Biology* 10:65–68.

Kuehler, C., and P. N. Witman. 1988. Artificial incubation of California Condor *(Gymnogyps californianus)* eggs removed from the wild. *Zoo Biology* 7:123–32.

Langston, W., Jr. 1981. Pterosaurs. *Scientific American* 244:122–36.

Latta, F. F. 1976. *The saga of Rancho el Tejón*. Santa Cruz, Calif.: Bear State Books.

Lawson, D. 1975. Pterosaur from the latest Cretaceous of West Texas: Discovery of the largest flying creature. *Science* 187: 947–48.

Lewis, M., and W. Clark. 1905. *Original journals of the Lewis and Clark expedition 1804–1806,* ed. R. G. Thwaites. Vol. 3, part 2; vol. 4, parts 1 and 2. New York: Dodd Mead and Co.

Ligon, J. D. 1967. Relationships of the cathartid vultures. *Occasional Papers, Museum of Zoology, University of Michigan* 651:1–26.

Linsdale, J. M. 1931. Facts concerning the use of thallium in California to poison rodents. . . . *Condor* 33:96–106.

Lint, K. C. 1943. Rearing an Andean Condor. *Zoonooz* 16(4):3.

Locke, L. N., G. E. Bagley, D. N. Frickie, and L. T. Young. 1969. Lead poisoning and aspergillosus in an Andean Condor. *Journal of the American Veterinary Medical Association* 155:1052–56.

Lyon, M. W., Jr. 1918. Occurrence of California Vulture in Idaho. *Journal of the Washington Academy of Sciences* 8:25–28.

Mallette, R. D., and J. C. Borneman. 1966. *First cooperative survey of the California Condor. California Fish and Game* 52(3): 185–203.

McMillan, I. 1953. Condors, politics, and game management. Reprint, *Central California Sportsman* 13(12):458–60.

———. 1968. *Man and the California Condor.* New York: Dutton.

———. 1970. Botching the condor program. *Defenders of Wildlife News* 45:95–98.

Meretsky, V. J., and N. F. R. Snyder. 1992. Range use and movements of California Condors. *Condor* 94:313–35.

Meretsky, V. J., N. F. R. Snyder, S. R. Beissinger, D. A. Clendenen, and J. W. Wiley. 2000. Demography of the California Condor: Implications for reestablishment. *Conservation Biology* 14: 957–67.

———. 2001. Quantity versus quality in California Condor reintroduction: Reply to Beres and Starfield. *Conservation Biology* 15:1449–51.

Mertz, D. B. 1971. The mathematical demography of the California Condor population. *American Naturalist* 105:437–53.

Miller, A. H. 1953. The case against trapping California Condors. *Audubon Magazine* 55:261–62.

Miller, A. H., I. McMillan, and E. McMillan. 1965. The current status and welfare of the California Condor. *National Audubon Society Research Report,* no. 6:1–61.

Miller, L. H. 1942. Succession in the cathartine dynasty. *Condor* 44:212–13.

Mundy, P. J. 1982. *The comparative biology of southern African vultures.* Johannesburg, South Africa: Vulture Study Group.

Mundy, P., D. Butchart, J. Ledger, and S. Piper. 1992. *The vultures of Africa.* London: Academic Press.

Mundy, P. J., and J. A. Ledger. 1976. Griffon vultures, carnivores and bones. *South African Journal of Science* 72:106–10.

———. 1977. The plight of the Cape Vulture. *Endangered Wildlife* 1:2–3.

Murphy, R. C. 1925. *Bird islands of Peru.* New York: Knickerbocker Press.

Nicolaus, L. K., J. F. Cassel, R. B. Carlson, and C. R. Gustavson. 1983. Taste aversion conditioning of crows to control predation on eggs. *Science* 220:212–14.

Olendorff, R. R., and R. N. Lehman. 1986. *Raptor collisions with utility lines: An analysis using subjective field observations.* Final report, U.S. Department of the Interior to Pacific Gas and Electric Company, San Ramon, California.

Owre, O. T., and P. O. Northington. 1961. Indication of the sense of smell in the Turkey Vulture from feeding tests. *American Midland Naturalist* 66:200–205.

Pattee, O. H., P. H. Bloom, J. M. Scott, and M. R. Smith. 1990. Lead hazards within the range of the California Condor. *Condor* 92:931–37.

Phillips, J. C. 1926. An attempt to list the extinct and vanishing birds of the Western Hemisphere, with some notes on recent status, locations of specimens, etc. *International Ornithological Congress* 6:503–34.

Plug, I. 1978. Collecting patterns of six species of vultures (Aves: Accipitridae). *Annals of the Transvaal Museum* 31:51–63.

Ralls, K., J. D. Ballou, B. A. Rideout, and R. Frankham. 2000. Genetic management of chondrodystrophy in California Condors. *Animal Conservation* 3:145–53.

Rea, A. M. 1983. Cathartid affinties: A brief overview. In *Vulture bi-*

ology and management, ed. S. R. Wilbur and J. A. Jackson, 26–54. Berkeley: University of California Press.

Richardson, P. R. K., P. J. Mundy, and I. Plug. 1986. Bone-crushing carnivores and their significance to osteodystrophy in griffon vulture chicks. *Journal of Zoology* (London) 210:23–43.

Ricklefs, R. E., ed. 1978. Report of the advisory panel on the California Condor. *National Audubon Society Conservation Report* 6:1–27.

Risebrough, R. W. 1986. Pesticides and bird populations. *Current Ornithology* 3:397–427.

Robinson, C. S. 1939. *Observations and notes on the California Condor from data collected on Los Padres National Forest.* U.S. Forest Service internal report, Santa Barbara.

———. 1940. *Notes on the California Condor, collected on Los Padres National Forest, California.* U.S. Forest Service internal report, Santa Barbara.

Sarrazin, F., C. Bagnolini, J. L. Pinna, and E. Danchin. 1996. Breeding biology during establishment of a reintroduced Griffon Vulture *Gyps fulvus* population. *Ibis* 138:315–25.

Sarrazin, F., C. Bagnolini, J. L. Pinna, E. Danchin, and J. Clobert. 1994. High survival estimates of Griffon Vultures *(Gyps fulvus fulvus)* in a reintroduced population. *Auk* 111:853–62.

Schoenherr, A. A. 1992. *A natural history of California.* Berkeley: University of California Press.

Scott, C. D. 1936a. Are condors extinct in Lower California? *Condor* 38:41–42.

———. 1936b. Who killed the condor? *Nature Magazine* 28(6): 368–70.

Sibley, F. C. 1968. The life history, ecology and management of the California Condor *(Gymnogyps californianus).* Annual Progress Report, Project No, B-22, U.S. Fish and Wildlife Service, Patuxent Wildlife Research Center, 34 pp.

———. 1969. *Effects of the Sespe Creek Project on the California Condor.* Laurel, Md.: U.S. Fish and Wildlife Service.

Simons, D. D. 1983. Interactions between California Condors and humans in prehistoric far western North America. In *Vulture biology and management,* ed. S. R. Wilbur and J. A. Jackson, 470–94. Berkeley: University of California Press.

Snyder, N. F. R. 1983. California Condor reproduction, past and present. *Bird Conservation* 1:67–86.

————. 1988. California Condor. In *Handbook of North American Birds,* ed. R. S. Palmer, 43–66. New Haven, Conn.: Yale University Press.

Snyder, N. F. R., S. R. Derrickson, S. R. Beissinger, J. W. Wiley, T. B. Smith, W. D. Toone, and B. Miller. 1996. Limitations of captive breeding in endangered species recovery. *Conservation Biology* 10:338–48.

Snyder, N. F. R., and J. A. Hamber. 1985. Replacement-clutching and annual nesting of California Condors. *Condor* 87:374–78.

Snyder, N. F. R., and E. V. Johnson. 1985. Photographic censusing of the 1982–1983 California Condor population. *Condor* 87:1–13.

Snyder, N. F. R., E. V. Johnson, and D. A. Clendenen. 1987. Primary molt of California Condors. *Condor* 89:468–85.

Snyder, N. F. R., and V. J. Meretsky. 2003. California Condors and DDE: A re-evaluation. *Ibis* 145:136–51.

Snyder, N. F. R., R. R. Ramey, and F. C. Sibley. 1986. Nest-site biology of the California Condor. *Condor* 88:228–41.

Snyder, N. F. R., and N. J. Schmitt. 2002. California Condor *(Gymnogyps californianus).* In *The Birds of North America,* no. 610, ed. A. Poole and F. Gill. Philadelphia, Pa.: The Birds of North America, Inc.

Snyder, N. F. R., and H. A. Snyder. 1989. Biology and conservation of the California Condor. *Current Ornithology* 6:175–267.

Snyder, N., and H. Snyder. 2000. *The California Condor, a saga of natural history and conservation.* London: Academic Press.

Snyder, N. F. R., and J. W. Wiley. 1976. Sexual size dimorphism and hawks and owls of North America. *Ornithological Monographs,* no. 20.

Stager, K. 1964. The role of olfaction in food location by the Turkey Vulture *(Cathartes aura). Los Angeles County Museum Contributions to Science* 81:1–63.

Steadman, D. W., and N. G. Miller. 1986. California Condor associated with spruce-jackpine woodland in the late Pleistocene of New York. *Quaternary Research* 28:415–26.

Stoms, D. M., F. W. Davis, C. B. Cogan, M. O. Painho, B. W. Duncan, and J. Scepan. 1993. Geographic analysis of California Condor sighting data. *Conservation Biology* 7:148–58.

Streator, C. P. 1888. Notes on the California Condor. *Oologist* 13(2):30.

Studer, C. D. 1983. Effects of Kern County cattle ranching on California Condor habitat. Master's thesis, Michigan State Universtiy, East Lansing.

Taylor, A. S. 1859a. The egg and young of the California Condor. *Hutchings's California Magazine* 3(12):537–40.

———. 1859b. The great condor of California. *Hutching's California Magazine* 3(12):540-43, 4(1):17–22, 4(2):61–64.

Terrasse, M. 1985. *Reintroduction du Vautour fauve dans les Grandes Causes (Cevennes).* Saint Cloud, France: Fonds d'Intervention pour les Rapaces.

Toone, W. D., and A. C. Risser. 1988. Captive management of the California Condor *(Gymnogyps californianus). International Zoo Yearbook* 27:50–58.

Toone, W. D., and M. P. Wallace. 1994. The extinction in the wild and reintroduction of the California Condor *(Gymnogyps californianus).* In *Creative conservation: Interactive management of wild and captive animals,* ed. P. J. S. Olney, G. M. Mace, and A. T. C. Feistner, 411–19. London: Chapman and Hall.

U.S. Fish and Wildlife Service. 1975. *California Condor recovery plan.* Approved April 9,1975, Washington, D.C., 63 pp.

———. 1996. *California Condor recovery plan,* 3d rev. Portland, Ore., 62 pp.

Verner, J. 1978. *California Condors: Status of the recovery effort.* General Technical Report PSW-28, U.S. Forest Service., Washington, D.C.

Wallace, M. P., and S. A. Temple. 1987a. Competitive interactions within and between species in a guild of avian scavengers. *Auk* 104:290–95.

———. 1987b. Releasing captive-reared Andean Condors to the wild. *Journal of Wildlife Management* 51:541–50.

Ward, J. C., and D. A. Spencer. 1947. Notes on the pharmacology of sodium fluoroacetate—compound 1080. *Journal of the American Pharmaceutical Assocociation* 36:59–62.

Wellnhoffer, P. 1991. *The illustrated encyclopedia of pterosaurs.* New York: Crescent Books.

Wetmore, A. 1933. The eagle—king of birds, and his kin. *National Geographic Magazine* 64(1):43–95.

Wiemeyer, S. N., R. M. Jurek, and J. R. Moore. 1986. Environmental contaminants in surrogates, foods, and feathers of California Condors *(Gymnogyps californianus). Environmental Monitoring and Assessment* 6:91–111.

Wiemeyer, S. N., A. J. Krynitsky, and S. R. Wilbur. 1983. Environmental contaminants in tissues, foods, and feces of California Condors. In *Vulture biology and management,* ed. S. R. Wilbur and J. A. Jackson, 427–39. Berkeley: University of California Press.

Wiemeyer, S. N., J. M. Scott, M. P. Anderson, P. H. Bloom, and C. J. Stafford. 1988. Environmental contaminants in California Condors. *Journal of Wildlife Management* 52:238–47.

Wilbur, S. R. 1972. *Food resources of the California Condor.* U.S. Fish and Wildlife Service, Patuxent Wildlife Research Center Administrative Report.

———. 1978a. Supplemental feeding of California Condors. In *Endangered birds, management techniques for preserving threatened species,* ed. S. A. Temple, 135–40. Madison: University of Wisconsin Press.

———. 1978b. The California Condor, 1966–1976: A look at its past and future. *U.S. Fish and Wildlife Service North American Fauna* 72:1–136.

———. 1980. Estimating the size and trend of the California Condor population, 1965–1978. *California Fish and Game* 66:40–48.

Wilbur, S. R., W. D. Carrier, and J. C. Borneman. 1974. Supplemental feeding program for California Condors. *Journal of Wildlife Management* 38:343–46.

ART CREDITS

VICTOR APANIUS (courtesy USFWS) plate 77

JOHN BORNEMAN (courtesy USFWS) plate 50

DAVE CLENDENEN (courtesy USFWS) plates 4, 28, 29, 41, 78, 86, 88, 90, 91

PHIL ENSLEY plate 11

WILLIAM FINLEY AND HERMAN BOHLMAN (courtesy Museum of Vertebrate Zoology, Berkeley) plates 32, 43, 44, 45

RON GARRISON (courtesy Zoological Society of San Diego) plates 67, 81, 82, 83, 84

JESSE GRANTHAM (courtesy USFWS) plate 56

ED HARRISON (courtesy USFWS) plate 47

JACK INGRAM (courtesy USFWS) plate 22

ERIC JOHNSON plate 57

CARL KOFORD (courtesy Museum of Vertebrate Zoology, Berkeley) plates 31, 35, 42, 69

CARL KOFORD (courtesy Western Foundation of Vertebrate Zoology) plate 53

ROBERT MCCABE (courtesy Museum of Vertebrate Zoology, Berkeley) plate 48

BILL NELSON maps 1, 2, 3, 4, 5

DAN PETERSON (courtesy USFWS) plate 54

JOHN SCHMITT plates 17, 33

HELEN SNYDER (courtesy USFWS) plates 1, 13, 27a, 39, 51, 58, 62, 65

INDEX

"A Doomed Bird," 96
Adams, M. S., 204–205
ADC. *See* Animal Damage Control
    (ADC)
African vultures, 17, 20, 153
    African White-backed Vulture,
        23 (plate)
    Bearded Vulture, 16, 27
    breeding, 150, 153
    calcium deficiency, 49–50, 49–51
        (plates), 51, 130–131
    Cape Vulture, 49–50, 49–50
        (plates), 51, 130
    griffon vultures, 59, 153
    Palm-nut Vulture, 21, 29
    poisoning, 118
African White-backed Vulture,
    23 (plate)
air-cell formation, 206
American Ornithologists' Union
    (AOU), 138, 187
ammunition
    "green bullet" program, 230–231
    ingested lead, 161–165, 161 (plate),
        222–223
    legal restrictions on lead, 213,
        229–233
    TTB-based, 229–233
Andean Condor
    captive breeding of, 182, 183, 190,
        191 (plate)
    similarities to California Condor,
        11, 12 (plate)
    studies, 159, 190, 191 (plate), 205

surrogate releases, 198, 214–218,
    214–215 (plates), 217 (plate),
    220 (plate)
Anderson, Marilyn, 166 (plate)
Animal Damage Control (ADC), 177
Antelope Valley, 101
Anthony, A. W., 116
Antilocapra americana, 54
AOU. *See* American Ornithologists'
    Union (AOU)
AOU-Audubon panel, 138, 187–188
Aquila chrysaetos. See Golden Eagle
Argentavis magnificens, 12–13, 15
"ark" philosophy, 210
Armadillo, 25 (plate)
artificial incubation, 192–196, 194
    (plate), 201–204, 201–202 (plates)
    See also captive breeding; replace-
        ment-clutching
Aspen Helicopters, 195 (plate)
aversive conditioning, 219, 220,
    220–221 (plates)

bacterial toxins, resistance to, 22, 29
Baja California, 108, 215, 216 (map)
Balaenoptera musculus, 2
Bald Eagle, 15, 167
Bancroft, H. H., 94
Barbour, Bruce, 157 (plate),
    178 (plate)
Bearded Vulture, 16, 27
bears, 73, 221
    Black Bear, 46, 65, 86
    Grizzly Bear, 34, 86, 116

behavioral problems, 203 (plate), 204,
210–211, 215, 217–218,
219 (plate), 224–225
combating, 219–222, 233–238
Beissinger, S. R., 163
Benchley, Belle, 183, 183 (plate)
Bendire, Charles, 97, 116, 120
Bidwell, J., 116
Bijleveld, M., 118
bill shape, relationship to diet, 15,
20–21, 23 (plate), 24, 25, 26
bismuth, 229
Bison, 59–60
Bison bison, 59–60
Bitter Creek National Wildlife Refuge,
175, 243
Black Bear, 46, 65
threats to eggs and nestlings, 86
Black Vulture, 11, 41, 47 (plate), 153
nests, 71
blinds, 64–65
bloodsucking bugs, 73
Bloom, Pete, 157 (plate)
Blue Whale, 2
Bobcat, 34
body size, 2, 8, 11–14, 12–13 (plates)
relationship to scavenging, 20–21
Bohlman, Herman, 97, 98 (plate)
Boise (Idaho), 200
bones, search for, 48–53
Borneman, John, 103, 177
Bos taurus, 34, 35, 96, 123, 131
Bourassa, Gene, 159
breeding
brooding, 78, 80 (fig.)
of captives, 192, 199–207, 201–203
(plates)
courtship, 66–71, 67–70 (plates)
effort, 122–125, 144–152, 151
(table), 154
egg laying, 72 (plate), 75–77
fledglings, 83–86, 99 (plate)
hatching, 77–78, 202 (plate)
incubation, 75, 76, 76 (plate), 77
nest attendance, 78, 80 (fig.), 81
nest sites, 3, 3 (plate), 36, 36 (plate),
54–56, 55 (map), 56 (plate),
71–75, 72 (plate), 76 (plate),

84 (plate), 86–89, 87 (plate),
134–135
nestlings, 56, 78–83, 79 (plate), 87,
87 (plate), 99 (plate), 102 (plate),
202–203 (plates)
predation threats, 86–91, 88–90
(plates), 153
rates of, 22, 26–27, 77, 122–125,
144–154, 223–225
of released birds, 223–225
status (1982–1986), 151 (table)
studies in the 1980s, 144–156, 148
(plate), 151 (table), 155 (plate),
156
success, 152–154, 223–225
See also captive breeding; nests
British Museum, 96
brooding, 78, 80 (fig.)
Brown, W. M., 129
Brown Pelican, 127
bullets. *See* ammunition
Bureau of Land Management, 138
Burro, 34
Bush, Brad, 148 (plate)

cages, 200, 201 (plate)
calcium
decline hypotheses, 130–131
needs, 167–168
search for, 35, 48–53
California Condor
characteristics of, 7–14, 9–10
(plates)
colors, 8, 9–10 (plates), 45 (plate),
66, 68
compared to Andean Condor, 11,
12 (plate)
taxonomy, 11
wings and flight, 8, 9 (plate), 67
(plate), 88–89 (plates), 140
(plate), 158 (plate)
See also scavenging; vocalizations
California Department of Fish and
Game, 101, 138, 183–184
California Fish and Game Commission (CFGC), 161–162, 184, 198
California Ground Squirrel, 35,
120–121, 177

California Polytechnic State University, 139
California Sea Lion, 35
California State Legislature, 184
California Vulture, 11
Camelops spp., 36
camels, 36
Camp, R. J., 90
Campbell, Kenneth, 15
candling, 200, 201 (plate)
Canis
    familiaris, 34
    latrans, 34, 117, 165, 176
    lupus, 116
cannon-netting, 159
Cape Vulture, 49 (plate)
    calcium deficiencies, 49–51,
        50 (plate), 130
Capra spp., 34
captive breeding, 4, 180–207
    artificial incubation, 192–196, 194
        (plate), 201, 201–202 (plates)
    behavioral problems, 204
    early-release proposal, 196–199
    establishment of flock, 188–196,
        189 (plate), 194–195 (plates)
    genetic concerns, 200, 204–207
    initiation of, 138, 182–188, 183
        (plate), 186 (plate)
    mate compatibility, 188, 192, 205
    opposition to, 182–188
    puppet-rearing, 202, 203 (plate),
        214–218, 221–222, 237
    rearing procedures, 200–204,
        201–203 (plates)
    replacement-clutching, 77,
        190–196, 194–195 (plates),
        201–204, 217, 237–238
    reproductive performance, 199–200
captive population. *See* captive breeding
Caracara plancus, 25
Carrizo Plains, 101, 175
Cathartes aura. See Turkey Vulture
Cathartidae. *See* Vulteridae
Cat, Domestic, 34
cats, sabre-toothed, 6
Cattle, 24 (plate), 34, 35 (plate),
    43 (plate), 96, 123, 131

cave art, 95 (plate)
censusing efforts, 100–106, 139–144
    See also population size
ceremonial skirt, 95 (plate)
Cervus elaphes nannodes, 34, 53–54
Cetacea, 34
CFGC. *See* California Fish and Game
    Commission (CFGC)
Channel Islands, 21
chaparral habitats, 131, 134
chelation therapy, 163, 222
Chemnick, L. G., 204–205
Chicken, Domestic, 47, 205–206
chicks, 77–83, 79 (plates), 82–84
    (plates), 86–87, 99 (plate),
    201–204, 202–203 (plates)
    feeding, 78–81
    fledging, 83–85
    preening, 81
    weaning, 85–86
Cholame, 107
chondrodystrophy, 206
Clark, A. J., 167
Coast Range, 100–101
Cochran, Bill, 159
Collins, Paul, 34, 35
collisions
    aversive conditioning to, 219, 220,
        220–221 (plates)
    due to lead poisoning, 168–169
    with overhead wires, 128–129,
        129–130 (plates), 165–166
    as source of mortality, 128, 129–130
        (plates), 165–166
Common Raven, 23–24, 24 (plate), 27,
    40–41, 153
    as threats, 86, 88–89 (plates), 89–91,
        153, 178–179, 178 (plate)
compound 1080, 119, 121, 177
Condor, Andean
    captive breeding of, 182, 183, 190,
        191 (plate)
    similarities to California Condor,
        11, 12 (plate)
    studies, 159, 190, 191 (plate), 205
    surrogate releases, 198, 214–218,
        214–215 (plates), 217 (plate),
        220 (plate)

Condor, California
    characteristics of, 7–14, 9–10
        (plates)
    colors, 8, 9–10 (plates), 45 (plate),
        66, 68
    compared to Andean Condor, 11,
        12 (plate)
    taxonomy, 11
    wings and flight, 8, 9 (plate),
        67 (plate), 88–89 (plates),
        140 (plate), 158 (plate)
    See also scavenging; vocalizations
Condor preserves, 131, 172–176,
        173–174 (plates), 175 (map), 176
        (plate), 241–243
Condor Recovery Team, 187, 234
    early-release proposal, 196–198
congenital deformities, 206
Connochaetes taurinus, 59
conservation efforts (historical),
        172–179
    habitat protection, 172–176,
        173–174 (plates), 175 (map),
        176 (plate), 241–242
    miscellaneous, 178–179
    proposals by Koford, Miller, and
        McMillan, 176–177
    supplemental feeding, 178
conservation efforts (modern),
        138–139
    captive breeding, 180–207
    costs of, 228
    future efforts, 226–240
    habitat protection, 174–175,
        175 (map), 242–243
    releases, 208–225
    research, 136–169
Cooper, James G., 96–97, 114, 116, 120
Cooper Ornithological Society, 96
Coragyps atratus, 11, 41, 71, 153
Corvidae, 23–24
Corvus corax. See Common Raven
courtship, 66–71
    mounting, 70–71
    mutual preening, 66, 67–68, 68–69
        (plates)
    pair flights, 66–67, 67 (plate)
    wing-out displays, 66, 68–71,
        70 (plate)

Cowles, R. B., 130–131
Coyote, 34, 117, 121, 165, 177
Crested Caracara, 25–26, 25 (plate)
Cretaceous, 12, 19
Crocuta crocuta, 17, 49, 130
crops
    adaptation for scavenging, 22, 26
    capacity of, 37
    distension of, 47
CV pair 1983, 147–149
cyanide poisoning, 119–120, 119
        (plate) 165, 166 (plate)
Cygnus buccinator, 167

Dasyhelea sp., 73
Dasypus novemcincius, 25 (plate)
Dawson, William Leon
    decline hypotheses, 114, 117
    population estimates, 97, 100, 107
DDE contamination, 127–128, 135,
        155–156, 155 (plate)
DDT. *See* DDE contamination
decline hypotheses
    calcium stress, 130–131
    censusing efforts, 100–106, 139–144
    collisions, 128–129, 129–130
        (plates), 165–166, 168–169
    Cooper's assessment of, 96–97
    DDE contamination, 127–128, 135,
        155–156, 155 (plate)
    disturbance of nesting areas,
        125–127, 154
    fire, 134–135, 134 (plate)
    food scarcity, 122–125, 135, 154
    habitat loss, 130–133, 154
    lead poisoning, 113, 160–162
        (plates), 161–165, 167–168, 169,
        211–213
    mortality studies, 156–167,
        157–158 (plates), 160–162
        (plates), 166 (plate), 167–169
    oil deposits, 133
    poisoning, 96–97, 116–122, 119
        (plate), 127–128, 135, 155–156,
        157, 160–162 (plates), 161–168,
        166 (plate), 167–169
    reproductive studies, 144–156, 148
        (plates), 151 (table), 155 (plate)
    sacrificial ceremonies, 133–134

shooting, 96–97, 114–116, 115 (plate), 135, 166
Deer, Mule, 34, 35, 53–54, 53 (plate), 60, 61, 131, 165
deer-hunting season, 59
diet
    calcium needs, 35, 48–53, 49–50 (plates), 130–131, 167–168
    diversity, 16–17, 21, 34–37, 35 (plate)
    See also calcium; food; foraging; scavenging
Dipodomys spp., 35, 120
division of labor between sexes, 22, 29–30, 77, 81
Dog, Domestic, 34
Domestic Cat, 34
Domestic Chicken, 47, 205–206
Domestic Dog, 34
domestic livestock, 60–61
Domestic Pig, 34
dominance relationships around carcasses, 42–46, 43–44 (plates)
double-clutching. *See* replacement-clutching

eagles. *See* Bald Eagle; Golden Eagle
early-release proposal, 196–199
Easton, Robert, 173
eating behavior
    of adults, 46–53, 47 (plate), 51–52 (plates)
    of chicks, 78–81, 82 (plate)
Eaton Canyon, 97, 98–99 (plates)
egg laying, 75–77
    See also replacement-clutching
eggs
    artificial incubation of, 192–196, 194–195 (plates)
    assisted hatching, 200–201, 202 (plates)
    candling, 200, 201 (plate)
    collection of, 126–127, 192–196, 194–195 (plates)
    malpositioning of embryos, 205–206
    mechanical shaking of, 193
    placement of, 72 (plate)
    protection of, 178–179

eggshell thinning, 127–128, 135, 155–156
Elephant Seal, Northern, 54
Elk, Tule, 34, 53
Emslie, Steve, 34, 36–37
Endangered Species Acts, 177, 187
epizootic bovine abortion, 59
Equus
    asinus, 34
    caballus, 34, 35, 131
    caballus × asinus, 34
Equus spp., 36
Eurasian Griffon Vulture, 118
evolution in captivity, 210
extinctions of wildlife, 6, 11–16, 19–20, 49–50, 228
eyesight, adaptation for scavenging, 22, 26

Falco
    mexicanus, 90–91
    peregrinus, 127
Falconidae, 25
falcons
    Peregrine Falcon, 127
    Prairie Falcon, 90–91
feathers
    collecting, 95 (plate), 116
    colors, 8, 9–10 (plates), 66
    gold dust containers, 116
    identification of individuals by, 105, 139–143
feeding chicks, 78–81, 82 (plate)
feet, adaptation for scavenging, 22, 24, 25, 26
Felis
    catus, 34
    concolor, 35, 46, 65
Finley, William, 97, 98–99 (plates)
    decline hypotheses, 114, 116
fire, 131, 134–135, 134 (plate)
fish, 18, 22
fledging, 83–86, 83–84 (plates)
fledgling production, 192–196, 199–200
food
    availability of, 60–61, 96–97, 122–125, 132, 154
    diversity of, 34–37

eating behavior, 46–53, 47 (plate),
    51–52 (plates)
quantitative needs, 37–38
regurgitation of food for young, 22,
    24, 29, 78–81, 82 (plate)
subsidy, 124–125, 162–163, 178,
    187, 212–217, 228–229
See also foraging; scavenging
foothill disease, 59
foraging
    after release, 162–163, 210,
        212–217, 228–229
    of breeders and nonbreeders, 55
        (map), 57–58
    and decline of native mammals,
        60–61
    eating behavior, 46–53, 47 (plate),
        51–52 (plates)
    finding and competing for food,
        38–46, 39 (plate), 43–45 (plates)
    regions, 53–57, 53 (plate), 55 (map),
        56 (plate), 175–176, 176 (plate)
    related to seasonal changes, 58–60,
        212–213
forested areas
    feeding in, 42
Fox, Gray, 35
Fry, D. M., 121, 163
Fry, W., 117

Gallus domesticus, 47, 205–206
genetic concerns, 200, 204–207
Geyer, C. J., 204
giant ground sloths, 6
giant sequoia, 2
    nest sites, 3, 3 (plate), 71, 75,
        148 (plate)
    wing-out display on, 70 (plate)
giants, 2–16, 37
    avian, 11–16, 14 (plate)
    flying reptiles, 11–15, 13 (plate)
    species in California, 2–3
gnats, 73
goats, 34
gold dust containers, 116
Golden Eagle, 15
    dominance of, 42–44, 43–44 (plates)
    lead contamination of, 164

threats to eggs and nestlings, 56–57,
    56 (plate), 86, 87–89, 88 (plate),
    132
Grand Canyon, 36, 36 (plate), 60, 215
Grantham, Jesse, 157 (plate)
Gray Fox, 35
Great Plains, 59–60
"green bullet" program, 230–231
griffon vultures, 15, 59, 153
    African White-backed Vulture, 23
        (plate)
    Cape Vulture, 49–50, 49–50
        (plates), 51, 130
    Eurasian Griffon Vulture, 118
Grinnell, Joseph, 96, 97, 100
Grizzly Bear
    as Condor food, 34
    poisoning campaigns, 116
    threats to nests, 86
ground sloths, giant, 6
Ground Squirrel, California, 35, 177
Gymnogyps californianus. See Cali-
    fornia Condor
Gypaetus barbatus, 16
Gypohierax angolensis, 21, 29
Gyps
    africanus, 23 (plate)
    coprotheres, 49–50, 130
    fulvus, 118

habitat
    as limiting factor, 131–133, 154
    loss, 131–133
    protection, 172–176, 241–243
Haliaeetus leucocephalus, 15, 167
Harris, Harry
    decline hypotheses, 117, 133
    Native American ceremonies, 94, 133
    population estimates, 96, 97
Harrison, E. N., 190
Hartt, E. W., 204, 236
hatching, 77–78
hawks, 11
head, naked, 22, 25, 26, 30
head color, 10 (plate), 45 (plate), 66, 68
Heerman, A. L., 133
Hegdahl, P. L., 121, 177
Henshaw, H. W., 116

Hi Mountain nesting area, 177
historic photographs, 79 (plate), 84 (plate), 95 (plate), 98–100 (plates), 102–103 (plates), 129 (plate), 173 (plate), 182 (plate)
hooked bills, 15, 20–22, 24, 25, 26, 30
Hopper Mountain National Wildlife Refuge, 175, 242
Hornaday, W. T., 60
Horse, 34, 35, 131
   native, 36
Houston, David, 16, 17, 18–19, 41, 42, 47, 59
Hudson Ranch. See Bitter Creek National Wildlife Refuge
humans
   artifacts, 48–53, 51–52 (plates), 223
   isolation of captive nestlings from, 200–201, 201 (plate)
   lead poisoning of, 168
   and Pleistocene extinctions, 37
   as stress to Condors, 125–127
hunting
   and lead poisoning, 229–233
   See also shooting
Hyena, Spotted, 17, 49, 130

identification system, 139–143, 140–142 (plates)
incubation, 75–77
Ingram, Jack, 148 (plate)
intelligence of scavengers, 22, 24, 25, 27–28

Johnson, Eric, 139, 142 (plate), 150
Jurassic, 19

Kahl, Phil, 38
kangaroo rats, 35, 120
Kern County, 108, 124, 141, 175–176
   foraging in, 55 (map), 58–59
Kiff, L. F., 127–128, 190
King Vulture, 11, 41
Knight, H. A. L., 90
Knight, R. L., 90
Knoder, Gene, 187
Koford, Carl, 102 (plate), 129 (plate)
   breeding studies, 64, 84, 102 (plate), 152, 153

conservation efforts, 173, 176–177
decline and stress hypotheses, 114, 116, 117, 120–121, 125, 128, 129 (plate), 131, 133
diet studies, 34, 35, 51, 61
flight studies, 38
nest monitoring, 84, 84 (plate), 87, 145, 146
opposition to captive breeding, 182–183
population estimates, 96, 100–102, 105, 106 (fig.), 107–108
support of nest sanctuaries, 125
Kuehler, Cyndi, 202 (plate)

La Brea tar deposits, 13, 133
Lammergeier. See Bearded Vulture
Lawson, Douglas, 12, 14
lead poisoning, 113, 160–162 (plates), 161–165, 167–169, 211
   ammunition fragments, 48, 161, 161 (plate)
   in humans, 168
   solutions to, 213, 228–233
Ledger, John, 49–51, 130
Ledig, Dave, 155 (plate)
Lehman, R. N., 129
Leptoptilos crumeniferus, 22–23, 38
Lepus californicus, 35, 131
Lewis and Clark expedition, 96
lice, 73
life span, 22, 66
Lint, K. C., 183
Lion, California Sea, 35
lions, 17
   Mountain Lion, 34, 46, 65
long neck, adaptation for scavenging, 22, 24–26, 30
Long-tailed Weasel, 35
Los Angeles County, 108
Los Angeles Zoo, 138, 182, 199–200
Los Padres National Forest, 100, 100 (plate), 177, 241–243
Lynx rufus, 34
Lyon, M. W., 116

maladaptive changes in captivity, 210
Mallette, R. D., 102

malpositioning of embryos, 205–206

mammals, near extermination of, 60

mammoths, 6, 36

Mammuthus spp., 6, 36

Marabou Stork, 22–23, 23 (plate), 25, 38

marine shells, 35

mate compatibility in captivity, 188, 192, 205

maturity
  age of, 66
  slow development of, 22, 26–27

Maurer, J. R., 163

McLean, D. D., 101

McMillan, Eben and Ian
  decline hypotheses, 114, 120–122, 126, 131, 135, 169
  diet studies, 60, 61, 125
  nest monitoring, 75
  opposition to captive breeding, 183
  population estimates, 102, 105, 106 (fig.), 107–108

mechanical shaking of eggs, 193

Megatherium spp., 6

mentor-rearing, 222

Menzies, Archibald, 96

Mephitis mephitis, 35

Meretsky, V. J., 143, 163, 167, 222

Mertz, D. B., 143

metabolic rate, 20–21, 37–38

Mexican government, 215

Miller, Alden, 103 (plate)
  breeding studies, 152
  decline hypotheses, 114, 120–122, 125, 131, 135, 169
  diet studies, 60, 61, 125
  opposition to captive breeding, 183, 184–185
  population estimates, 102, 103 (plate), 105

mineral deposits in Sespe region, 174

Miocene, 19

Mirounga angustirostris, 54

molt, 38
  patterns, 139–141, 140 (plate)

Monterey County, 108
  foraging range in 1940s, 60

mortality rates
  as cause of decline, 143–144, 156–169
  determination through photos, 143
  in early-release program, 196–197
  lack of age dependence, 144
  of released birds, 222–223

Mountain Lion, 34, 46, 62

movements
  of breeders and nonbreeders, 57–58
  between roosts and foraging, 54–57, 55 (map)
  and seasonal changes, 58–60

Mule, 34

Mule Deer, 34, 35, 53–54, 53 (plate), 60, 61, 131, 165

multiple-clutching. *See* replacement-clutching

Mundy, Peter
  breeding studies, 150, 153
  calcium deficiency studies, 49–50, 130
  strychnine poisoning studies, 118
  teratorn diet, 15–16
  vocalization studies, 28–29

museum specimen collection, 114, 126–127

Museum of Vertebrate Zoology, 97

Mustela frenata, 35

mutual preening, 66, 67–68, 68–69 (plates)

naked head, adaptation for scavenging, 22, 25, 26, 30

naked vulture of California, 8

name origin for California Condor, 8, 11

NAS. *See* National Audubon Society (NAS)

National Audubon Society (NAS)
  on captive breeding, 183, 187–188
  conservation actions, 177
  opposing capture, 162, 198–199
  radiotelemetry program, 157
  research support, 101, 102, 138

National Geographic Society, 102

National Park Service, 174–175, 175 (map)

Native American societies, 94,
    95 (plate), 133–134, 199
naturalistic releases, 219–220, 234–238
neck, adaptations for scavenging, 14,
    22, 24, 25, 26, 30
Neotoma spp., 72
nesting areas, 55 (map), 175 (map)
    human disturbance of, 125–127
    protection of, 172–174, 241–243
nests
    attendance during breeding period,
        78–80, 80 (fig.)
    breeding cages, 200–201, 201 (plate)
    characteristics of, 71–73
    diet remains in, 34–37
    investigations by adult Condors,
        73–75
    locations, 3, 3 (plate), 36, 36 (plate),
        54–56, 55 (map), 56 (plate),
        71–73, 72 (plate), 76 (plate),
        84 (plate), 86–87, 87 (plate)
    parasite infestations in, 73
    predation threats to, 73, 86–91,
        89–90 (plates), 153
    safety considerations, 86–91, 87–90
        (plates)
Nicolaus, L. K., 179
nontoxic ammunition, 165, 229–233
Northern Elephant Seal, 54

observational procedures for nests,
    64–65
October Survey, 102–104, 105,
    106 (fig.), 123, 185
Odocoileus hemionus, 34, 131, 165
Ogden, John, 157, 158 (plate)
oil deposits, 133
Olendorff, R. R., 129
Oligocene, 19
Oncorhynchus spp., 34
Orange County, 108
Oregon Zoo, 182
Ovis aries, 34, 35, 96

Pacific Ocean shore, foraging site, 53–54
pair flights, 66–67, 67 (plate)
pair formation, 66–71
    mounting, 70–71

mutual preening, 66, 67–68, 68–69
    (plates)
pair flights, 66–67, 67 (plate)
wing-out displays, 66, 68–71,
    70 (plate)
Palm-nut Vulture, 21, 29
Panama, 47
Panthera leo, 17
parasite infestations in nests, 73
parent-rearing, vs. puppet-rearing,
    221–222, 233–238
parrots, 21
Pattee, O. H., 164
Patuxent Wildlife Research Center,
    190, 191 (plate)
pecking order, 44–46
Pelecanus occidentalis, 127
Pelican, Brown, 127
Percy Ranch. *See* Hopper Mountain
    National Wildlife Refuge
Peregrine Falcon, 127
Peregrine Fund, 182, 200, 215
Peru, Andean Condor releases, 214
pesticides, 127–128, 155–156
photo census, 105, 106, 106 (fig.),
    140–143, 140–142 (plates), 150
photographs, historic, 79 (plate), 84
    (plate), 95 (plate), 98–100
    (plates), 102–103 (plates), 129
    (plate), 173 (plate), 182 (plate)
pictograph, 94, 95 (plate)
Pig, Domestic, 34
Pine Mountain, 177
Pine Mountain Club, 218, 219 (plate)
Piru Gorge, 177
pit traps, 159
Pleistocene
    foraging, 19–20, 36–37, 60
    geographical range during, 4, 6,
        6 (map)
    nests, 36–37, 36 (plate), 71
    size of Condors during, 13–14
Plunkett, Dick, 187
poisoning, from predator control,
    96–97, 116–122, 119 (plate),
    166 (plate)
population size, 106 (fig.)
    captive, 188–200

early records, 94–97
early twentieth century, 106–109
first estimates, 97–104
1980s, 105–106, 139–144
Prairie Falcon, 90–91, 90 (plate)
predation threats
    Common Ravens, 86, 88–89 (plates),
        89–91, 153, 178–179, 178 (plate)
    to eggs and nestlings, 86–91, 88–90
        (plates)
    Golden Eagles, 56–57, 56 (plate),
        86, 87–89, 88 (plate), 132
    Prairie Falcons, 90–91, 90 (plate)
predators, poisoning campaigns
        against, 116–122
preening, 38, 81, 82 (plate)
    mutual, 66, 67–68, 68–69 (plates)
Pronghorn, 54
protected areas. See sanctuaries
Pteronodon longiceps, 11, 13 (plate), 18
pterosaurs, 11–12, 13 (plate), 14–15,
        17–18
    diet, 17–18
puppet-rearing
    declining value, 237–238
    vs. parent-rearing, 221–222
    procedures, 202, 203 (plate)
    release results, 214–219
    See also captive breeding

Quetzalcoatlus sp., 14–15, 18, 19
    northropi, 12, 14–15, 18

rabbits, 35, 131
radiotelemetry, 130, 157–160, 158
        (plate), 160 (plate), 191 (plate)
Ralls, Katherine, 206
Ramey, Rob, 194 (plate)
range, historical and prehistorical,
        4–7, 5–6 (maps), 36–37,
        36 (plate), 53–55, 53 (plate),
        55 (map)
rats, kangaroo, 35, 120
ravens. See Common Raven
regurgitation of food for young
    adaptation for scavenging, 22,
        24–25, 29
    in Condors, 80, 82 (plate)

releases
    areas of, 215, 216 (map)
    behavioral problems, 217–222, 217
        (plate), 219–221 (plates)
    breeding problems, 223–225
    collision deaths, 129, 130 (plate),
        219, 220, 220–221 (plates)
    integration into wild populations,
        215
    lead poisoning, 162–163, 222–223,
        225
    methods and results, 214–225,
        214–215 (plates), 216 (map),
        217–221 (plates)
    naturalistic, 219, 234–238
    strategies, 233–238
    surrogate Andean Condors,
        214–217, 214–215 (plates),
        217 (plate), 220 (plate)
replacement-clutching, 77, 190–196,
        194–195 (plates), 201–204, 217,
        237–238
    penalties of, 237–238
reproduction. See breeding; captive
        breeding
reproductive studies of the 1980s,
        144–156
resistance to bacterial toxins, 22, 29
Ricklefs, Robert, 188
Risebrough, Robert, 128
Risser, Art, 166 (plate), 189 (plate)
Riverside County, 108
Robinson, Cyril, 100, 100 (plate), 107,
        173
Royal Vulture, 11

sabre-toothed cats, 6
sacrificial ceremonies, 133–134
salmon, 34
San Benito County, 108
San Diego County, 108
San Diego Wild Animal Park, 199–200,
        199 (plate), 201 (plate)
San Diego Zoological Society, 138,
        162 (plate), 182, 183, 185, 187,
        189 (plate), 190, 194–195 (plates),
        199, 199 (plate), 200, 201–203
        (plates)

San Emigdio Ranch. *See* Wind Wolves
    Preserve
San Joaquin Valley, 102, 154
    decline of livestock, 61
    foraging sites, 53 (plate), 55 (map),
        176 (plate)
San Luis Obispo County, 107, 108
    foraging range in 1940s, 60
sanctuaries
    creation of, 172–176, 173–174
        (plates), 175 (map), 176 (plate),
        241–243
    U.S. Forest Service pledge to pro-
        tect, 197
    See also Sespe Condor Sanctuary;
        Sisquoc Condor Sanctuary
Santa Barbara County, 100–101, 108,
    173, 177
Santa Clara County, 108
Santa Cruz County, 108
Sarcoramphus papa, 11, 41
scavenging
    adaptations for, 20–31
    lifestyle, 16–20
    by odor, 41–42, 41 (plate)
    Old World and New World com-
        pared, 42
    relationship of body size and bill
        shape to, 20–21
    risks and problems of, 40–46, 43–44
        (plates)
Scheuhammer, A. M., 167
Schmitt, John, 45 (plate), 82 (plate),
    160 (plate)
Scott, C. D., 114, 116, 117
Sea Lion, California, 35
Seal, Northern Elephant, 54
sequoia, giant, 2
    nest sites, 3, 3 (plate), 71, 75,
        148 (plate)
    wing-out display on, 70 (plate)
Sequoiadendron giganteum. See
    sequoia, giant
Serengeti Plain, 17, 59
Sespe Condor Sanctuary, 100–101,
    108, 124, 148 (plate), 175 (map)
    establishment of, 173–174, 174
        (plate), 241–243

firearms bans in and near, 177, 242
releases into, 214–219, 214–215
    (plates), 216 (map), 217–218
    (plates)
sexual dimorphism, lack of, associated
    with scavenging, 22, 30
Sheep, 34, 35, 96
shells, marine, 35
shot pellets, 165, 230
shooting, 96–97, 114–116, 115 (plate),
    135, 166
Sibley, Fred
    breeding studies, 64, 153
    decline hypotheses, 124, 169
    nest-monitoring, 87, 126, 145
    population estimates, 103–104, 104
        (plate), 105, 106, 106 (fig.), 107
Sierra Madre Ridge, 177
Sierra Nevada, 148 (plate)
Sierra San Pedro Martir, 108
Simons, D. D., 134
Sisquoc Condor Sanctuary, 100–101,
    108, 173 (plate), 175 (map)
    establishment of, 173, 241
size, adaptation for scavenging, 20, 21
Skunk, Striped, 35
sloths, giant ground, 6
smell, finding food by, 41–42
Smilodon spp., 6
Snedden, Bert, 102
soaring efficiency, 19
sodium fluoroacetate, 119
Spencer, D. A., 121
Spermophilus beecheyi, 35, 120–121,
    177
Spotted Hyena, 17, 49, 130
squirrels, 120–121
    California Ground Squirrel, 35,
        120–121, 177
storks, 11, 12
    adaptation for scavenging, 22–23
    Marabou Stork, 22–23, 23 (plate),
        25, 38
straight bills, 22, 23
Streator, C. P., 116
Striped Skunk, 35
strychnine, 116–120
Styrofoam objects, 52, 52 (plate)

Suckley, G., 96
 sunning activities, 38, 39 (plate)
supplemental food, 124–125,
    162–163, 178, 187, 212–217,
    228–229
Sus scrofa, 34
Swan, Trumpeter, 167
Sylvilagus audubonii, 35, 131

Tapir, 47 (plate)
Tapirus bairdii, 47 (plate)
Taylor, A. S., 96, 116
Tehachapi foothills, 58
Tejón Ranch region, 53 (plate),
    118–119, 120, 124, 175–176
Teratornis
    incrediblis, 13
    merriami, 13–15, 14 (plate)
teratorns, 12–15
thallium, 121
The California Condor: A Saga of Nat-
    ural History and Conservation, 4
The Sign, 141, 141 (plate)
thermals, 38, 40
ticks, 73
tin, 229
tongue, structure of, 48
Tonni, Eduardo, 15
Toone, Bill, 189 (plate)
Topatopa, 185, 186 (plate), 205
    finding a mate for, 188–190, 192
Transverse Range, 100–101
trapping free-flying birds, 159
trash, parental feeding of, 48–49,
    50 (plate), 223–224
Truesdale, Kelly, 146
Trumpeter Swan, 167
TTB ammunition, 229–233
Tule Elk, 34, 53–54
tungsten, 229
Turkey Vulture, 11, 17, 27, 40–41,
    41 (plate)
    breeding, 71, 153
    eating behavior, 47
    lead contamination of, 164
    nests, 71
    olfactory abilities, 41–42
    resistance to compound 1080, 121

Urocyon cinereoargenteus, 35
Ursus
    americanus, 46, 65, 86
    arctos horribilis, 34, 86, 116
U.S. Congress, 138, 188
U.S. Fish and Wildlife Service
    (USFWS)
    on captive breeding, 187, 188
    on capture of wild birds, 161–162,
        198
    on combating behavioral problems,
        219–222
    release programs, 215, 234
    research, 103–104, 126, 138, 177
U.S. Forest Service
    protected areas, 174, 175 (map),
        197, 241–243
    research support, 138
U.S. Military, conversion to nontoxic
    ammunitions, 230–231
USFWS. See U.S. Fish and Wildlife
    Service (USFWS)
utility poles, 129, 130 (plate), 165–166,
    219–220, 220–221 (plates)

Ventana Wilderness Society, 215,
    221–222
Ventura County, 100–101, 108, 123,
    124, 173, 177
Verner, Jared, 103, 143, 187
vocalizations, 8, 22, 25, 28, 30
Vulteridae, 11
Vultur gryphus. See Andean Condor
Vulture Study Group, 159
vultures, 8, 11
    African vultures, 17, 20, 118, 150,
        153
    African White-backed Vulture,
        23 (plate)
    Andean Condor, 11, 12 (plate), 159,
        182, 183, 190, 191 (plate), 197,
        198, 214–217, 217 (plate), 220
        (plate)
    Bearded Vulture, 16, 27
    Black Vulture, 11, 41, 47 (plate), 71,
        153
    California Condor as, 11
    California Vulture, 11

Cape Vulture, 49–50, 49–50
    (plates), 51, 130
Eurasian Griffon Vulture, 118
griffon vultures, 15, 59, 153
King Vulture, 11, 41
Old World vultures, 11, 17, 20–21,
    42, 59
Palm-nut Vulture, 21, 29
Royal Vulture, 11
scavenging lifestyle, 16–31
Turkey Vulture, 11, 17, 27, 40–42, 41
    (plate), 47, 71, 121, 153, 164

Wallace, Mike, 159, 214–215
Ward, J. C., 121
weaning, 85–86
Weasel, Long-tailed, 35
West Nile virus, 238
Westermark effect, 236
whales, 34
    Blue Whale, 2
White-backed Vulture, African,
    23 (plate)
Wiemeyer, S. N., 164
Wilbur, Sanford
    breeding studies, 154

decline hypotheses, 114, 117,
    122–125, 132, 135
food studies, 61, 122–125, 132
population estimates, 104,
    104 (plate)
recovery plan, 187
Wildebeest, 59
Wind Wolves Preserve, 175, 176
    (plate)
winds and thermals, 38–39, 47–48
wing-out displays, 66, 68–71,
    70 (plate)
wing span, 8
wolves, 116
woodrats, 72

Xolxol, 189 (plate)

Zalophus californianus, 35
zoo environments
    breeding in, 223, 225
    diseases in, 183, 184
    See also captive breeding; Los Ange-
        les Zoo; San Diego Zoological
        Society

Series Design:	Barbara Jellow
Design Enhancements:	Beth Hansen
Design Development:	Jane Tenenbaum
Illustrator/Cartographer:	Bill Nelson
Composition:	Jane Rundell
Text:	9.5/12 Minion
Display:	ITC Franklin Gothic Book and Demi
Printer and Binder:	Everbest Printing Company

ABOUT THE AUTHORS

The professional careers of Noel and Helen Snyder have been devoted mainly to field conservation studies of various endangered birds, with special emphasis on the Puerto Rican Parrot, the Thick-billed Parrot, the Florida Everglade Kite, and the California Condor. The Snyders have also worked with other wildlife species, however, including aquatic snails, and have devoted considerable time to writing and music. Noel earned simultaneous bachelor's degrees in music (majoring in cello) at the Curtis Institute of Music and in biology at Swarthmore College, and he completed a Ph.D. in evolutionary biology at Cornell University in 1966. Helen completed her bachelor's degree in biology at Oberlin College and was also pursuing graduate studies at Cornell when they began their long-standing collaborations.

The Snyders' field efforts have included behavioral and ecological studies of many raptor species, resulting in numerous scientific papers and a popular book on the raptors of North America, published by Voyageur Press in 1992. In 1972, they joined the Endangered Wildlife Research program of the U.S. Fish and Wildlife Service and the U.S. Forest Service to conduct research on the Puerto Rican Parrot, and this effort was followed by studies of other endangered birds for the federal and various state governments and the National Audubon Society. For work on the Puerto Rican Parrot and California Condor, Noel received the William Brewster Award of the American Ornithologists' Union, a distinguished achievement award from the Society for Conservation Biology, and the Conservation Medal of the Zoological Society of San Diego.

From the late 1980s to the present, the Snyders have been based in southern Arizona, where they have conducted research on Goshawks and Thick-billed Parrots. These activities have been interspersed with participation in research and conservation training programs focused on parrots in Jamaica, St. Lucia, and Mexico for Wildlife Preservation Trust International. They played an important role in recent successful efforts to ensure the protection of Cave Creek Canyon of the Chiricahua Mountains from mining development and have spent considerable recent energies aiding campaigns to preserve important montane habitats in Mexico.